AF358623

Marine Biologically Active Compounds as Feed Additives

Marine Biologically Active Compounds as Feed Additives

Editor

Izabela Michalak

MDPI • Basel • Beijing • Wuhan • Barcelona • Belgrade • Manchester • Tokyo • Cluj • Tianjin

Editor
Izabela Michalak
Wrocław University of Science and Technology
Poland

Editorial Office
MDPI
St. Alban-Anlage 66
4052 Basel, Switzerland

This is a reprint of articles from the Special Issue published online in the open access journal *Journal of Marine Science and Engineering* (ISSN 2077-1312) (available at: https://www.mdpi.com/journal/jmse/special_issues/feedadditives).

For citation purposes, cite each article independently as indicated on the article page online and as indicated below:

LastName, A.A.; LastName, B.B.; LastName, C.C. Article Title. *Journal Name* **Year**, *Article Number*, Page Range.

ISBN 978-3-03943-470-1 (Hbk)
ISBN 978-3-03943-471-8 (PDF)

Cover image courtesy of Izabela Michalak.

Contents

About the Editor

Izabela Michalak holds a master's degree in Biotechnology (2005) and a Ph.D. degree in Chemical Technology, specialization in Biotechnological processes, from Wrocław University of Science and Technology (2010). She is currently Associate Professor at the same University, a post she has held since her appointment in 2019. She has authored more than 100 peer-reviewed papers in international journals and book chapters in addition co-editing two books ("Algae Biomass: Characteristics and Applications: Towards Algae-Based Products. Series Title: Developments in Applied Phycology", Springer, and "Innovative Bio-Products for Agriculture: Algal Extracts in Products for Humans, Animals and Plants", Nova Science Publishers). Her research interests concern biosorption of metal ions by seaweeds (wastewater treatment and production of feed additives), extraction of active compounds, and application of algal products in agriculture.

Preface to "Marine Biologically Active Compounds as Feed Additives"

The marine environment consists of a wide variety of organisms with beneficial properties. Among them, a special role is played by macroalgae (seaweeds), which are amongst the first multicellular organisms and, as such, the precursors to land plants. The growing scope of seaweed-based applications in food, agricultural fertilizers, animal feed additives, pharmaceuticals, cosmetics, and personal care is expected to boost market demand. Agriculture and animal feed applications have held the second largest seaweed market share in 2017, and the combined market is anticipated to reach much higher values by 2024 due to the impacts of current research and development targeting enhanced animal health and productivity.

Seaweeds have a long tradition of being used in animal feed, especially in coastal areas. They are a rich source of biologically active compounds (pigments, proteins, amino acids, phlorotannins, polyunsaturated fatty acids, vitamins, and carbohydrates such as agar, alginate, and carrageenan) and minerals (iodine, zinc, sodium, calcium, manganese, iron, selenium) and are thus considered as natural feed additives. In most cases, seaweeds are mixed with animal feed because when consumed alone, they can have a negative impact on animals. The nutritional value of seaweeds and their effect on different species of animals were described in the reviews of Tiago Morais et al., Garima Kulshreshtha et al., Izabela Michalak and Khalid Mahrose, and Melania L. Cornish et al. Tiago Morais et al. presented, in detail, seaweeds as a valuable nutritional and nutraceutical animal feed additive, including fish and oyster farming, poultry (laying hen and broiler chickens), and in ruminant feed with an emphasis on the reduction in methane emissions from ruminants. In this Special Issue, particular attention was paid to animal health. In the reviews of Garima Kulshreshtha et al. and Izabela Michalak and Khalid Mahrose, seaweeds as sustainable feed sources for poultry health and production were discussed. The effect of seaweed-supplemented diets on growth, performance, gastrointestinal flora, disease, immunity, and overall health of laying/broiler hens was presented. Melania L. Cornish et al. highlighted the extensive prebiotic effects of selected macroalgae. Due to their unique properties, seaweeds can serve as an alternative to antibiotic growth promoters. In the research article of Garima Kulshreshtha et al., it was shown that red seaweeds *Chondrus crispus* and *Sarcodiotheca gaudichaudii* and their selected, purified components can be used to increase the lifetime of existing, patented antibiotics and can also help in reducing the costly (economic and environmental) therapeutic and prophylactic use of antibiotics in poultry. Red seaweeds have demonstrated antimicrobial properties against the poultry pathogen *Salmonella enteritidis*. An interesting aspect was raised by Izabela Michalak and Khalid Mahrose – inclusion of seaweeds in animal feed can enrich animal-derived products with active compounds, such as micro- and macroelements, polyunsaturated fatty acids, and pigments, and decrease the content of cholesterol. Michalak Izabela et al. tested the effect of green macroalga *Enteromorpha* sp. enriched with Zn(II) and Cu(II) ions on the daily amounts of feces and urine excreted by growing pigs, apparent fecal nutrient digestibility, and daily nitrogen balance and retention, meat quality, and the slaughter value of carcasses. It was suggested that *Enteromorpha* sp. may be introduced into pig nutrition as a feed material providing an alternative to inorganic salts due to enrichment of meat with microelements, proteins, decrease in fat content, lower water absorption and drip loss from meat, and a slight darkening of meat. Seaweeds have the potential to be commonly used as feed additives not only thanks to their properties but also due to the fact that the search for new, cheaper, safe feed additives is a priority of animal husbandry.

I would like to thank all the contributors for their hard work, commitment, and enthusiasm, which made it possible to accomplish this Special Issue.

Izabela Michalak
Editor

Journal of
*Marine Science
and Engineering*

Article

Antimicrobial Effects of Selected, Cultivated Red Seaweeds and Their Components in Combination with Tetracycline, against Poultry Pathogen *Salmonella* Enteritidis

Garima Kulshreshtha [1,2], Alan Critchley [3], Bruce Rathgeber [4], Glenn Stratton [1], Arjun H. Banskota [5], Jeff Hafting [6] and Balakrishnan Prithiviraj [1,2,*]

[1]   Department of Plant, Food, and Environmental Sciences, Agricultural Campus, Dalhousie University, P.O. Box 550, Truro, NS B2N 5E3, Canada; GR784654@DAL.CA (G.K.); g.stratton@dal.ca (G.S.)

[2]   Department of Cellular and Molecular Medicine, Faculty of Medicine, University of Ottawa, Ottawa, ON K1H 8M5, Canada

[3]   Verschuren Centre for Sustainability in Energy and Environment, Cape Breton University, Sydney, Cape Breton, NS B1P 6L2, Canada; alan.critchley2016@gmail.com

[4]   Department of Animal Science and Aquaculture, Agriculture Campus, Dalhousie University, P.O. Box 550, Truro, NS B2N 5E3, Canada; brathgeber@dal.ca

[5]   Aquatic and Crop Resource Development, National Research Council Canada, 1411 Oxford Street, Halifax, NS B3H 3Z1, Canada; Arjun.Banskota@nrc-cnrc.gc.ca

[6]   Acadian Seaplants Limited, 30 Brown Avenue, Dartmouth, NS B3B 1X8 Canada; jhafting@acadian.ca

*   Correspondence: bprithiviraj@dal.ca; Tel.: +1-902-893-6643; Fax: +1902-895-6734

Received: 24 June 2020; Accepted: 10 July 2020; Published: 12 July 2020

Abstract: Poultry and its products are an economical source of high-quality protein for human consumption. In animal agriculture, antibiotics are used as therapeutic agents to treat disease in livestock, or as prophylactics to prevent disease and in so doing enhance production. However, the extensive use of antibiotics in livestock husbandry has come at the cost of increasingly drug-resistant bacterial pathogens. This highlights an urgent need to find effective alternatives to be used to treat infections, particularly in poultry and especially caused by drug-resistant *Salmonella* strains. In this study, we describe the combined effect of extracts of the red seaweeds *Chondrus crispus* (CC) and *Sarcodiotheca gaudichaudii* (SG) and compounds isolated from these in combinations with industry standard antibiotics (i.e., tetracycline and streptomycin) against *Salmonella* Enteritidis. Streptomycin exhibited the higher antimicrobial activity against *S.* Enteritidis, as compared to tetracycline with a MIC_{25} and MIC_{50} of 1.00 and 1.63 µg/mL, respectively. The addition of a water extract of CC at a concentration of 200 µg/mL in addition to tetracycline significantly enhanced the antibacterial activity (log CFU/mL 4.7 and 4.5 at MIC_{25} and MIC_{50}, respectively). SG water extract, at 400 and 800 µg/mL ($p = 0.05$, $n = 9$), also in combination with tetracycline, showed complete inhibition of bacterial growth. Combinations of floridoside (a purified red seaweed component) and tetracycline (MIC_{25} and MIC_{50}) in vitro revealed that only the lower concentration (i.e., 15 µg/mL) of floridoside potentiated the activity of tetracycline. Sub-lethal concentrations of tetracycline (MIC_{50} and MIC_{25}), in combination with floridoside, exhibited antimicrobial activities that were comparable to full-strength tetracycline (23 µg/mL). Furthermore, the relative transcript levels of efflux-related genes of *S.* Enteritidis, namely *marA*, *arcB* and *ramA*, were significantly repressed by the combined treatment of floridoside and tetracycline, as compared to control MIC treatments (MIC_{25} and MIC_{50}). Taken together, these findings demonstrated that the red seaweeds CC and SG and their selected, purified components can be used to increase the lifetime of existing, patented antibiotics and can also help to reduce costly (economic and environmental) therapeutic and prophylactic use of antibiotics in poultry. To our knowledge, this is the first report of antibiotic potentiation of existing industry standard antibiotics using red seaweeds and their selected extracts against *S.* Enteritidis.

Keywords: red seaweeds; floridoside; antibiotics; efflux pumps; *Salmonella*; poultry

---

1. Introduction

Poultry and their products are an economic source of high-quality protein for human consumption. A range of feed additives including antibiotics, phytogenics or phytobiotics, probiotics and prebiotics, have been used by the poultry industry in order to improve both feed efficiencies and also the health and productivity of layer hens and broilers [1–3]. In livestock, the use of antibiotics for growth promotion was phased out in Canada [4] but is widely used in many parts of the world.

Despite these developments, it is currently estimated that over 60% of all antibiotics produced are used in livestock production, including poultry [5–7]. In 2012, it was estimated that 14.6 million kg of antibiotics were sold for use in animal agriculture [8], which was four times (3.29 million kg) the amount of antibiotics used for human use [9]. Currently, commercial poultry farms have higher rearing densities and the scale of production has dramatically increased to meet consumer demand. This has increased the frequency of outbreaks of infectious disease within flocks and therefore disease outbreaks which has required further interventions with antibiotics. In North America, antibiotics including chlortetracycline, lincomycin, oxytetracycline, penicillin, tylosin and virginiamycin are approved for use in poultry [4,10]. Antibiotics exert their effect by reducing the colonization of bacteria, increasing the metabolism of beneficial bacteria and reducing the total load of bacteria in the gut, thus reducing the overall bacterial load [11]. Sub-therapeutic levels of antibiotics also enhance immune responses of the host to an invading pathogen. Roura et al. (1992) showed that inclusion of streptomycin and penicillin in the diets of chicks resulted in preventing immunological stress by lowering cytokines [12].

However, the overuse of antibiotics in livestock came at a cost of increasing numbers of drug-resistant, bacterial pathogens. In 1951, Starr and Reynolds first reported a case of antibiotic resistance in bacteria in turkeys. The use of streptomycin as a growth promoter in turkey poults resulted in drug-resistant coliforms within three days of application [13]. In 1994, sixty-two isolates of vancomycin-resistant *Enterococcus faecium* were obtained from non-human sources in the United Kingdom (UK), amongst which 22 were from farm animals. This indicated that farm animals served as a reservoir for the development of drug-resistant bacteria [14]. Following this report, avoparcin was the first antibiotic to be banned in Europe in 1995. Consequently, the European Union (EU) banned the use of antibiotic growth-promoters in 2006 [15]. The selection pressure caused by antibiotics on gut microbes resulted in the development of resistant genes, which are transferred amongst species of pathogenic bacteria by horizontal gene transfer. This resulted in the excessive growth of resistant bacterial pathogens such as *Clostridium*, *Salmonella* and *Campylobacter* in the host, resulting in harmful diseases. In addition, changes in the microbial population within the gut can make the host more vulnerable to infections by other environmental pathogens [16].

In the United States, the Food and Drug Administration (FDA) controls the use of cephalosporin in animal agriculture. Also, there is increased interest to exclude the use of fluoroquinolones and tetracyclines in animal production. This is because these antibiotics are commonly used in treating bacterial infection in humans. In the EU and North America there is a heightened public awareness of the negative effects of antibiotics in livestock production. Therefore, there is increasing interest to develop alternatives to antibiotics [17]. Other control measures, such as competitive exclusion and vaccination, have contributed significantly to reduce pathogen (especially *Salmonella*) infections in layer production [18]. According to the U.S. Centers for Disease Control and Prevention, every year more than 2.8 million humans are infected with antibiotic-resistant bacteria, which leads to approximately 35,000 deaths [19]. It is clear that drug-resistance in pathogenic bacteria has developed since the middle of the last century, an era when antibiotics were used extensively to treat both human and animal diseases. It is likely that the emergence of drug-resistant strains of pathogenic bacteria is due to the flagrant large-scale overuse of antibiotics in medicine and agriculture [20].

Bacteria acquire antibiotic resistance by several mechanisms, including (i) drug inactivation/ modification, (ii) alteration of the target site (iii), bypass pathways and (iv) decreased membrane permeability. In addition, antibiotic resistance develops due to formation of biofilms and the inactivation of antibiotics by bacterial enzymes, modification in the outer membrane lipid bi-layer and porin permeability and sequestration of antibiotics within the bacterial biofilms [21–24]. Therefore, there is an urgent need to find effective alternatives that can be used to treat infections caused by drug-resistant *Salmonella* strains in humans and farm animals.

Some antimicrobial therapies involve the use of antimicrobial peptides, cell membrane permeabilizers, molecular chaperones, DNA synthesis and efflux-pump inhibitors. However, despite being effective in in-vitro studies, none of these strategies have advanced to clinical trials [25]. An alternative approach to finding new antibiotic classes is to potentiate the activity of already existing, registered/patented antibiotics using combined therapies. Several antimicrobial peptides, molecules, plant extracts and essential oils have been shown to enhance the activity of antibiotics, such as chloramphenicol, ciprofloxacin and tetracycline against Gram-positive and Gram-negative bacteria [26,27].

Tetracyclines are broad-spectrum bacteriostatic antibiotics that interfere with protein translation by inhibiting the attachment of aminoacyl-tRNA to the ribosomal acceptor (A) site. Tetracycline forms a complex with Mg^{2+} and blocks aminoacyl-tRNA binding and thus inhibits protein synthesis [28]. Essential oils from *Salvia* species (Lamiaceae) have been shown to potentiate the efficacy of tetracycline by inhibiting efflux pumps in *Staphylococcus epidermis*. The inhibition of the Tet (K) efflux pump of tetracycline resistant *S. epidermidis* by essential oils from three salvia species [29]. Moreover, organic extracts of pomegranate, myrrh and thyme significantly increased the efficacy of tetracycline against both Gram-positive and Gram-negative pathogens. This suggested that combinations with natural compounds could be used to enhance the efficacy of "fading" antibiotics [26].

Floridoside 2-*O*-α-D-galactopyranosylglycerol is a neutral heteroside found in red algae. It plays an important role in osmotic acclimation and provides resistance to osmotic stress in red algae [30]. Floridoside also has potent medicinal properties and has been shown to possess anti-viral and antitumor activities [31]. Earlier, Khan et al. (2012) reported alginate, a polysaccharide found in brown seaweeds, potentiated the antimicrobial activity of antibiotics against pathogens such as *Pseudomonas*, *Acinetobacter* and *Burkholderia* spp. [25]. Here, we describe the combined effects of selected extracts of two red seaweeds, i.e., *Chondrus crispus* and *Sarcodiotheca gaudichaudii*, and along with two well-used antibiotics (i.e., tetracycline and streptomycin) against *Salmonella* Enteritidis.

2. Materials and Methods

2.1. Bacterial Strain, Chemicals and Antibiotics

Nalidixic acid-resistant *Salmonella* Enteritidis was provided by the Laboratory for Foodborne Zoonoses, Public Health Agency of Canada, Guelph, Ontario. Half strength tryptic soy agar (TSA) medium (Difco) supplemented with nalidixic acid (32 µg/mL) was used for bacterial growth [32,33]. The antibiotic discs (BBL™ Sensi-Disc™), of tetracycline (TE30; 30 µg), streptomycin (S10; 10 µg), erythromycin (E15; 15 µg), novobiocin (NB30; 30 µg), penicillin (P10; 30 µg) and triple sulfa (SSS25; 15 µg) were purchased from Becton (BBL™ Sensi-Disc™), Dickinson and Company Franklin Lakes, NJ, USA. Acadian Seaplants Limited, kindly donated the two seaweeds which were cultivated on land in Charlesville, Nova Scotia, Canada. The extracts were prepared as described previously by Kulshreshtha et al. (2016) [34]. Tetracycline and streptomycin were obtained from Sigma Aldrich (Oakville, ON, Canada). Stock solutions of antibiotics and seaweed extracts were prepared and stored at −20 °C. Other chemicals and media used in this study were purchased from Difco Laboratories, Baltimore, MD, USA.

2.2. Antibiotic Sensitivity Assay

Susceptibility of *S.* Enteritidis to antibiotics was determined using the disc diffusion method, as described by the Clinical and Laboratory Standards Institute (CLSI) with some modifications [32,33]. Briefly, the bacterial culture (OD_{600} = 0.1, 1×10^8 cells/mL) was spread on a tryptic soy agar plate, before placing the antibiotic discs. Plates were incubated at 37 °C for 16–18 h and the diameter of the zone of growth inhibition was measured. The diameter of the paper disc was subtracted giving the growth-free zone of bacterial inhibition.

2.3. Determination of MIC of Antibiotics

The susceptibility of *S.* Enteritidis to the antibiotics tetracycline and streptomycin was tested by a broth inoculation method [32,33]. The testing of MICs (MIC_{25} and MIC_{50}) was performed in triplicate with an inoculum of 1×10^8 cells/mL. MICs were determined as the lowest concentration of antibiotics required for complete inhibition of bacteria after incubation at 37 °C for 16–18 h in an incubator shaking at 200 rpm. The MATLAB R2010a (curve fitting tool) was used to determine minimum inhibitory concentrations (MIC_{25} and MIC_{50}) of the antibiotics.

2.4. Combined Effect of Seaweed Extracts (SWE) and Antibiotics on Salmonella Enteritidis

The combined effect of extracts of *C. crispus* and *S. gaudichaudii* and antibiotics (tetracycline and streptomycin at MIC_{25} and MIC_{50}) were evaluated in-vitro using a broth inoculation method as described previously by Kulshreshtha et al. [34]. To 10 mL of tryptic soy broth, seaweed extract (SWE) and 100 µL *Salmonella* Enteritidis (OD_{600} = 0.1, 1×10^8 cells/mL) were added so that the final concentrations of SWE in 10 mL with tryptic soy broth were 200, 400, 800 µg/mL. Culture tubes were incubated at 37 °C for 24 h. The growth of *S.* Enteritidis was determined by plating the serially diluted culture on TSA plates to enumerate the colony forming units (CFU).

2.5. Extraction of Seaweed and Isolation of Floridoside

Water extracts of both seaweeds (SWE) were prepared as described previously by Kulshreshtha et al. (2016) for antibacterial test [34]. The proton nuclear magnetic resonance (^{1}H NMR) spectra of SWE were measured on a Bruker Advance III spectrometer (Bruker Biospin, Switzerland) operating at 700 MHz spectrometer with deuterated water to characterize major component. One of the major component of SWE, i.e., floridoside, was further purified from 80% EtOH extract, as shown in Scheme 1. Other seaweed components, including isethionic acid, citrulline and taurine were commercially obtained to test for their antibacterial activity.

2.6. Antimicrobial Effects of Seaweed Components on Salmonella Enteritidis

Floridoside, isethionic acid and taurine were identified in both CC- and SG-SWE extracts (Figure S1). L-Citrulline was also detected in SWE of *C. crispus*. Pure compounds (i.e., isethionic acid, taurine, L-Citrulline and floridoside) were tested in-vitro against *S.* Enteritidis by the broth inoculation method, as described in above Section 2.4. Fifteen µg/mL of pure compound was added to TSA broth and inoculated with *S.* Enteritidis. Antimicrobial activity was determined as a measure of log CFU/mL.

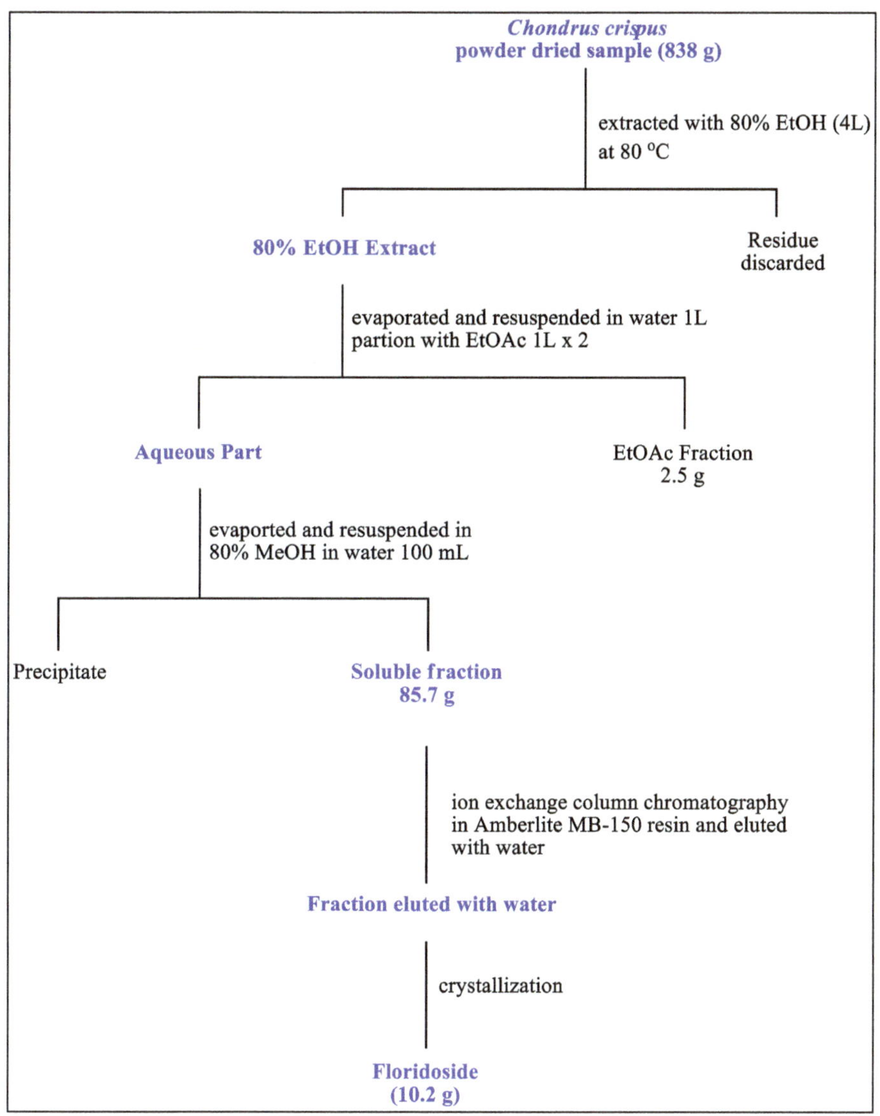

Scheme 1. Extraction and purification process of floridoside from 80% EtOH extract of *C. crispus*.

2.7. Combined Effects of Floridoside and Tetracycline on Salmonella Enteritidis

Synergistic interactions of floridoside and tetracycline (MIC_{25} and MIC_{50}) were evaluated in-vitro using the liquid culture inhibition test, as described in above Section 2.4. Briefly, bacterial cells were grown in the presence of different combination of floridoside (15 µg/mL) + tetracycline (MIC_{25}, 4 µg/mL), floridoside (15 µg/mL) + Tetracycline (MIC_{50}, 7.9 µg/mL). Tetracycline (MIC_{25} and MIC_{50}) and floridoside (15 µg/mL) were used as controls. Antimicrobial activity was determined as a measure of log CFU/mL.

2.8. Effects of Floridoside and Tetracycline on the Expression of Efflux-Pump-Related Genes

Gene expression analysis was carried out at time intervals of 45, 90 and 180 min to understand the mechanism of the combined effects of tetracycline and SWE. Briefly, bacterial cells from different treatments were centrifuged at 12,000× *g* for 10 min and total RNA was extracted using Trizol (Invitrogen), as described by the manufacturer. The RNA quality was assessed by agarose gel electrophoresis and quantified by NanoDrop ND-2000 spectrophotometer (NanoDrop Technologies Wilmington, DE). The relative transcript abundance of multi-drug efflux-pump genes were quantified using the StepOne Plus Real time PCR system (Applied Biosystems, ON, Canada), as described previously by Kulshreshtha et al. (2016) [34]. The gene specific primers used for this experiment are listed in Table 1. *16SrRNA* and *tufA* genes were used as internal control and the relative expression levels were calculated using the ΔΔCt method.

Table 1. The efflux-pump-related genes and primer sequences used in RT-qPCR.

Gene	Primer Sequence (5′ → 3′)
ramA	*CGTCATGCGGGGTATTCCAAGTG* *CGCGCCGCCAGTTTTAGC*
marA	*ATCCGCAGCCGTAAAATGAC* *TGGTTCAGCGGCAGCATATA*
acrB	*TTTTGCAGGGCGCGGTCAGAATAC* *TGCGGTGCCCAGCTCAACGAT*
16SrRNA	*GCGGCAGGCCTAACACAT* *GCAAGAGGCCCGAACGTC*
tufA	*TGTTCCGCAAACTGCTGGACG* *ATGGTGCCCGGCTTAGCCAGTA*

2.9. Statistical Analyses

A completely randomized design was followed for all assays. The experiments were performed three times, each with three biological replicates. Data were analyzed using ANOVA one-way analysis of variance with a *p* value of 0.05 using the statistical software Minitab and SAS. Log transformation was applied to the non-homogenous data before analysis. If significant main effects were found with ANOVA, the Tukey's procedure was used to compare differences among the least-square means. The standard deviation (SD) was reported with the mean. Differences were considered significant when *p* was <0.05.

3. Results

3.1. Screening of Antibiotics against Salmonella Enteritidis

The efficacy of antibiotics against *S.* Enteritidis was determined by the disc diffusion method via determination of the zone of growth inhibition. The antibiotics tetracycline, streptomycin, penicillin, erythromycin, triple sulfa and novobiocin were tested against *S.* Enteritidis. Amongst the antibiotics tested, tetracycline (30.0 μg) and streptomycin (10.0 μg) exhibited zones of inhibition of 22.5 and 18.0 mm, respectively). On the basis of the zone of inhibition interpretation chart, tetracycline and streptomycin were chosen for further studies.

3.2. Determination of Minimum Inhibitory Concentrations (MIC$_{25}$ and MIC$_{50}$)

The minimum inhibitory concentrations (MIC$_{25}$ and MIC$_{50}$) of the selected antibiotics (tetracycline and streptomycin) were determined using the MATLAB curve-fitting tool. For tetracycline, an MIC for 50% of the strain (MIC$_{50}$) was 4 μg/mL and 25% of the strains (MIC$_{25}$) was 7.9 μg/mL. Streptomycin exhibited a higher antimicrobial activity against *S.* Enteritidis, as compared to tetracycline with an MIC_{25} and MIC_{50} of 1 and 1.63 µg/mL, respectively.

3.3. SWE Potentiated the Effect of Antibiotics on Salmonella Enteritidis

The combined effects of SWE (both CC and SG), with antibiotics, was determined by a liquid culture inhibition test. Antibiotics (tetracycline and streptomycin) at MIC_{50} and MIC_{25} were combined with 200, 400, 800 µg/mL SWE (SG and CC) (Figure 1). The combination of tetracycline and CC at 400 µg/mL (log CFU 5.4 at MIC_{50}, $p = 0.01$, $n = 9$) and 800 µg/mL (log CFU 6.1 at MIC_{25} and 5.8 at MIC_{50}, $p = 0.01$, $n = 9$) did not affect the growth of *S.* Enteritidis, as compared to the tetracycline alone (log CFU 6.1 and 5.5 at MIC_{25} and MIC_{50} respectively, $p = 0.01$, $n = 9$). However, the combination of tetracycline at MIC_{25} and 400 µg/mL of CC-SWE were effective in reducing *S.* Enteritidis growth. Moreover, the lowest concentration of CC-SWE (200 µg/mL) and tetracycline (MIC_{25} and MIC_{50}) were the most effective in reducing bacterial growth (log CFU 4.7 and 4.5 at MIC_{25} and MIC_{50}, respectively) (Figure 1a). For SG-SWE, the response was dose-dependent, e.g., the higher concentration of SG-SWE (800 µg/mL, $p = 0.05$, $n = 9$) in combination with tetracycline showed complete inhibition of bacterial growth (Figure 1b). With 200 µg/mL of SG SWE bacterial growth was significantly reduced (log CFU 4.8 and 4.5 at MIC_{25} and MIC_{50}, respectively), compared to the MIC controls (log CFU 5.5) (Figure 1b). The antimicrobial effects of the SWE (both CC and SG) and streptomycin (MIC_{25} and MIC_{50}) were similarly tested. Trends were observed for streptomycin and SWE (CC and SG) against *S.* Enteritidis (Figure 1c,d). The combination treatments with the lowest concentration of CC-SWE (200 µg/mL, log CFU 4.1 and 4.3 at MIC_{50} and MIC_{25}, respectively, $p = 0.05$, $n = 9$) and the higher concentration of SG-SWE (800 µg/mL, log CFU 0 at MIC_{50} and MIC_{25}, respectively, $p = 0.05$, $n = 9$) were found to be the most effective (Figure 1c,d). In a comparison to the inhibitory effects of both antibiotic combinations with SWE, tetracycline showed the best combined effects and was used in further experiments.

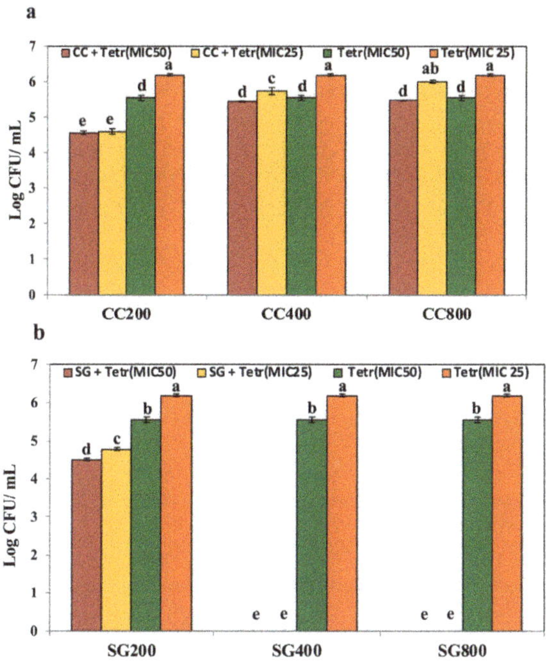

Figure 1. *Cont.*

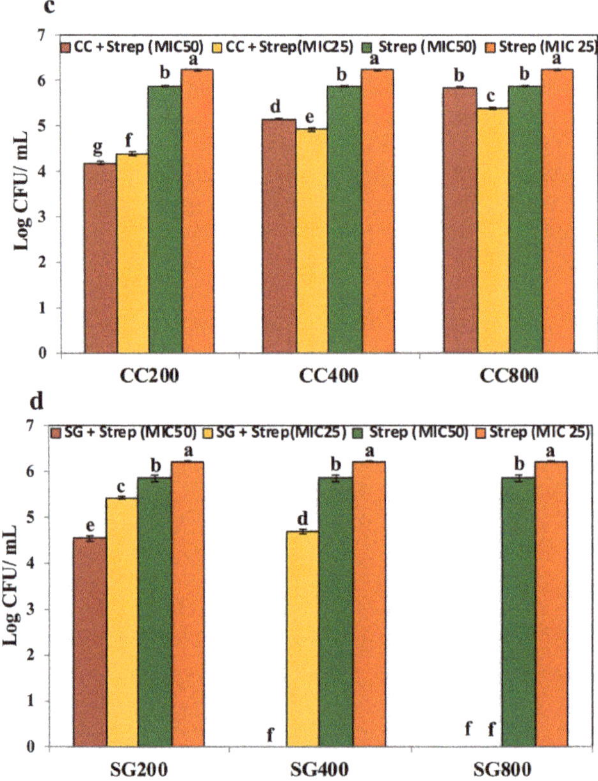

Figure 1. Combined effects of antibiotics and seaweed extracts (SWE) on *S*. Enteritidis. Tetracycline (Tetr) and streptomycin (Strep) at MIC_{50} and MIC_{25} were tested in combination with seaweeds *Chondrus crispus* (CC) and *Sarcodiotheca gaudichaudii* (SG) at three different concentrations (200, 400 and 800 µg/mL) (**a**) CC and Tetr; (**b**) SG and Tetr; (**c**) CC and streptomycin (Strep); (**d**) SG and Strep. Values with different superscript letters were significantly different ($p < 0.05$). Values represented mean ± standard deviation from three independent experiments ($n = 9$).

3.4. ^1H Nuclear Magnetic Resonance Spectroscopy of Seaweed Water Extracts

The NMR analysis identified three major compounds, namely isethionic acid, taurine and floridoside, in the water extracts of CC and SG (Figure S1). The ^1H NMR spectrum of floridoside isolated from 80% EtOH extract of *C. crispus* is shown in Figure S2.

3.5. Floridoside Affected the Growth of S. Enteritidis

The susceptibility of *S*. Enteritidis to purified seaweed compounds (i.e., isethionic acid, citrulline, taurine and floridoside) were tested using the liquid culture method. Floridoside and isethionic acid (15 µg/mL) reduced the colony count (log CFU/mL 6.21 and 6.33, respectively, $p = 0.09$, $n = 9$), as compared to control (log CFU/mL 6.5, $p = 0.09$, $n = 9$). However, higher colony counts (Log CFU/mL) of *S*. Enteritidis were observed on treatment with citrulline and taurine (Figure 2). Of the two most effective seaweed compounds (i.e., floridoside and isethionic acid), floridoside showed the highest antimicrobial activity and was selected for further evaluation.

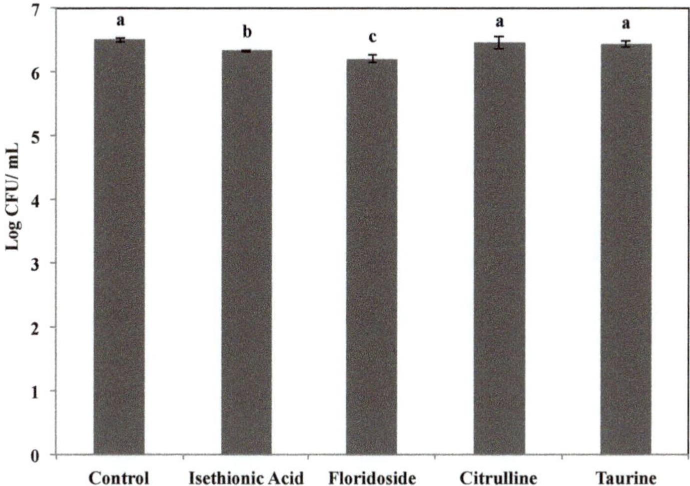

Figure 2. Antimicrobial effects of pure compounds from seaweed water extract (SWE) of CC on the growth of *S.* Enteritidis. Values with different superscript letters were significantly different ($p < 0.05$). Values represented mean ± standard deviation from three independent experiments ($n = 9$).

3.6. Floridoside Potentiated the Activity of Tetracycline against S. Enteritidis

Different concentrations of floridoside (15–100 μg/mL) in combination with tetracycline (MIC_{25} and MIC_{50}) were tested for their antimicrobial activity using a broth dilution method (Figure 3). Floridoside at 15 μg/mL potentiated the activity of tetracycline at both MICs (log CFU 4.3–5.2, $p < 0.05$, $n = 9$). Sub-lethal concentrations of tetracycline (MIC_{50} and MIC_{25}; 4 and 7.9 μg/mL, respectively) in combination with floridoside (15 μg/mL) exhibited antimicrobial activity which was comparable to full strength tetracycline (23 μg/mL). Compared to MICs alone, the combination of tetracycline (MIC_{25} and MIC_{50}) and 25 μg/mL of floridoside inhibited the growth (log CFU/mL 6.05 and 4.7, $p < 0.05$, $n = 9$) of *S.* Enteritidis (Figure 3). The number of bacterial aggregates, at higher concentrations of floridoside (i.e., 50 and 100 μg/mL), in combination with tetracycline, were not significantly different from the control ($p > 0.05$, $n = 9$).

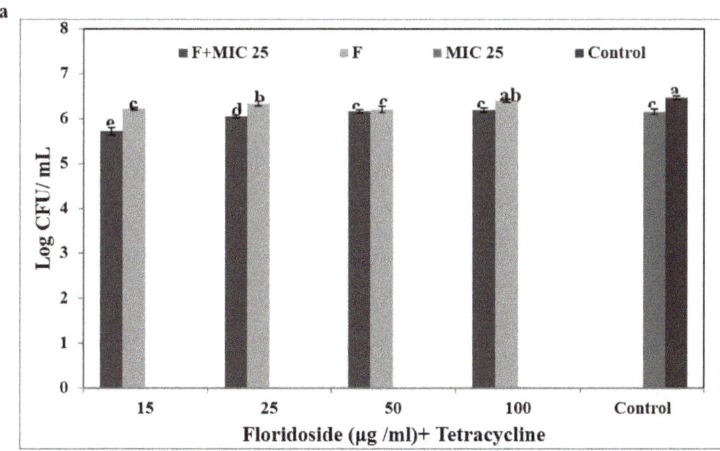

Figure 3. *Cont.*

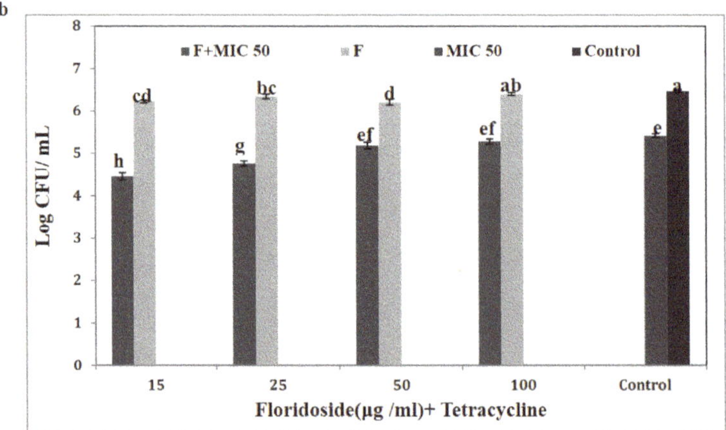

Figure 3. Combined effects of floridoside and tetracycline on the growth of *S*. Enteritidis. (**a**) MIC$_{25}$ (**b**) MIC$_{50}$. Values with different superscript letters were significantly different ($p < 0.05$). Values represented mean ± standard deviation from three independent experiments ($n = 9$). F + MIC25: combination of floridoside and tetracycline at MIC$_{25}$; F: floridoside; MIC 25: tetracycline at MIC$_{25}$; F + MIC50: combinations of floridoside and tetracycline at MIC$_{50}$. MIC 50: tetracycline at MIC$_{50}$.

3.7. Floridoside and Tetracycline Suppressed the Expression of Efflux-Pump-Related Genes

Gene expression analysis was conducted to understand the inhibitory mechanism of the combined effect of tetracycline and floridoside on *S*. Enteritidis. Real-time PCR analysis showed that the combination of floridoside and tetracycline (MIC$_{25}$ and MIC$_{50}$) suppressed the expression of efflux-related genes after 90 min of treatment (Figure 4). The relative transcript level of *marA*, which encodes a global regulator of multi-drug efflux-pumps was repressed by 2–15-fold, as compared to the control MIC treatments (Figure 4). Similarly, the *arcB* gene encoding the transporter component of the main efflux-pump (AcrAB) and *ramA*, a transcriptional activator of protein *ramA* involved in multi-drug efflux-pumps, were down-regulated by 18–25 fold and 14–20 fold, respectively ($p < 0.001$, $n = 9$) (Figure 4). This indicated that floridoside might favor the accumulation of tetracycline in the cell by repressing the expression of efflux-pump genes.

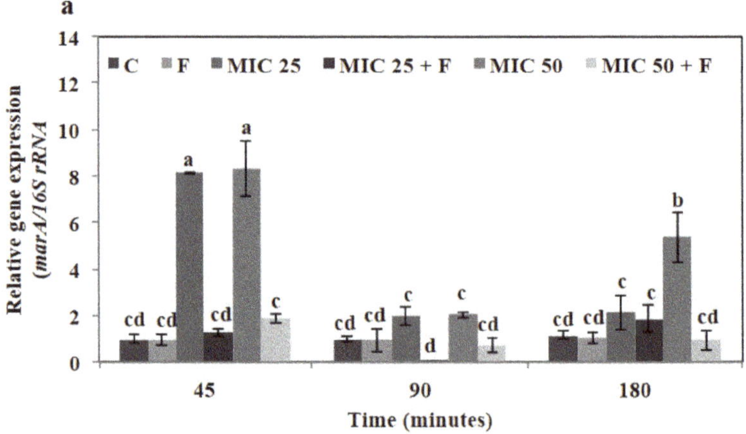

Figure 4. *Cont.*

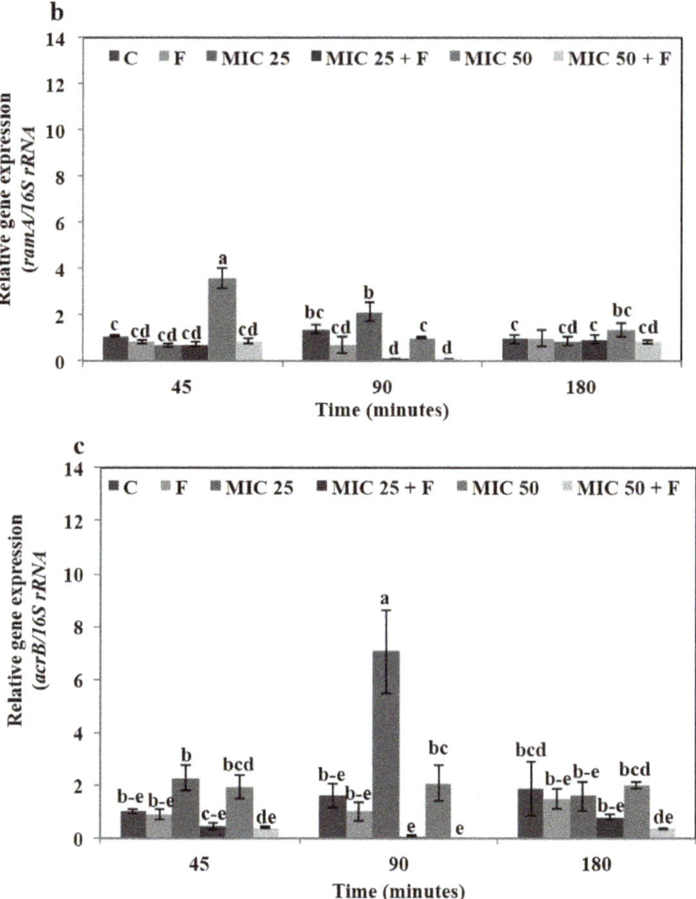

Figure 4. Effect of floridoside (F) on the expression of efflux-pumps related genes of *S*. Enteritidis. (**a**) *marA* (**b**) *ramB* (**c**) *acrA* after 45, 90 and 180 min of treatment with floridoside (15 µg/mL and Tetracycline (MIC$_{25}$ and MIC$_{50}$, 4 and 7.9 µg/mL)). Values with different superscript letters were significantly different ($p < 0.05$). C: control; F: floridoside; MIC 25: tetracycline at MIC$_{25}$; MIC25 + F: combination of tetracycline at MIC$_{25}$ and floridoside; MIC 50: tetracycline at MIC$_{50}$; MIC50 + F: combinations of tetracycline at MIC$_{50}$ and floridoside. Values represented mean ± standard deviation from three independent experiments ($n = 9$).

4. Discussion

Antimicrobials used in food animals contribute in the selection and dissemination of drug-resistant zoonotic, food-borne pathogens such as *Salmonella* Enteritidis. Non-typhoid *Salmonella* has become resistant to drugs, including ampicillin, chloramphenicol, quinolones and sulphonamide [35]. The main aim of the present study was to identify compounds from two specific red seaweeds (CC and SG) that improved the efficacy of existing, commercial antibiotics, in order to reduce their therapeutic and prophylactic use in poultry.

Microbes have utilized their innate genetic resistance and lateral gene transfer to acquire resistance to several antibiotics used in clinical and agricultural practices. This indicates that since the time antibiotics have first been employed, their success was compromised by over-usage and development of tolerance or resistance. Thus, in the era of diminishing activity of available antibiotics, an additive effect such as using combined therapies could enhance the life time of existing antibiotics [26,27].

Bacterial resistance to antimicrobial drugs can be related to their ability to form biofilms and secrete virulence factors. Previously, it has been shown that, among other functions, the matrix of biofilms prevents the access of antibiotics to the bacterial cells by sequestering them in the periplasm [24]. Furthermore, several studies also indicated the co-selection of virulence traits with antimicrobial drug resistance by integration of virulence and resistance plasmids. Up-regulation of virulence improves the fitness of the pathogen and has been shown to contribute to drug resistance [36]. In the present study, we determined the ability of SWE (from CC and SG) to potentiate the activity of existing antibiotics (i.e., tetracycline and streptomycin) by using well-established broth dilution and MIC assays. We observed that combinations of CC and SG water extracts with antibiotics significantly reduced the growth of *S.* Enteritidis by 3–6 fold, as compared to the antibiotics alone (Figure 1). Previously, we showed that SWE (CC and SG) reduced biofilm formation and down-regulated virulence gene expression of *S.* Enteritidis [34]. Therefore, the increase in bacterial susceptibility to antibiotics was most likely due to the effect of SWE (from both CC and SG) on biofilm formation and secreted virulence factors. This finding is beneficial as there is a lack in the discovery of new antibiotics. According to the most recent report from the Infectious Disease Society of America (ISDA), the numbers of antibiotics approved by FDA for marketing or in late-stage clinical development in the US has increased since IDSA's 2013 update [37]. However, a major concern is that a majority of these approved agents have been developed by modification of existing chemical classes of antibiotics, rather than new chemical classes. More importantly, large pharmaceutical sponsors continue to abandon the field and ISDA predicts that a sustainable antibiotics production will be bleak without further economic incentives for antibiotic development. Also, these drugs in the development pipeline might not be approved by the FDA and are not guaranteed to work against resistant human pathogens. It is suggested that new approaches to therapeutics other than small-molecule antibiotics that target resistant bacterial pathogens are desirable [38]. Hence, the current finding of the ability to revive ineffective doses of antibiotics by using natural, seaweed-derived compounds such as floridoside could be a suitable alternative. Currently, due to reduced financial incentives, pharmaceutical companies have limited their research on the development of new antibiotics. In this scenario, an alternative strategy to increase the efficacy of existing antibiotics could save the cost of production and development of new antibiotics. Thus, the implementation of combined therapies, i.e., the use of compounds such as floridoside, might improve existing antibiotic performance.

Floridoside is a neutral heteroside isolated from red algae and serves as a soluble carbon reserve. Floridoside from red seaweeds has also been researched for its potential medicinal and pharmaceutical applications. Park et al. (2007) [39] isolated floridoside from the red alga *Ahnfeltiopsis flabelliformis* and discovered its anti-quorum sensing activity. They identified that a mixture of seaweed compounds containing betonicine, floridoside and isethionic acid was capable of inhibiting AHL signaling in the quorum-sensing inhibition assay [39]. A year later, the same research group isolated the individual compounds and tested their effects on cell growth and quorum sensing using a reporter strain *Agrobacterium tumefaciens*. They observed that although the isolated floridoside had no effect on cell growth and quorum sensing, its combination with other isolated seaweed compounds significantly inhibited AHL activity [40]. In another study, Janssens et al. (2008) tested the effects of some red seaweed-derived compounds, i.e., furanones with tetracycline on the viable cell count of *Salmonella* biofilms. They concluded that pre-treatment of furanones reduced the viable cells in *Salmonella* biofilms by 50–2,100-fold [41]. Synthetic brominated furanones (Z)-4-bromo-5-(bromomethylene)-3-methylfuran-2(5H)-one (BF8) have been demonstrated to revert the antibiotic tolerate of *Pseudomonas aeruginosa* PAO1 persister cells. Treatment with BF8 at growth non-inhibitory concentrations (0.1–2 µg/mL) increased the susceptibility of persister cells to ciprofloxacin (Cip). Interestingly, BF8 was effective against both planktonic and biofilm forms of *P. aeruginosa* PAO1 [42]. In another study, a novel

3-chloro-5(S)-[(1R,2S,5R)-2-isopropyl-5-methylcyclohexyloxy]-4-[4-methylphenylsulfonyl]-2(5H)-furanone (F105) increases the efficacy of aminoglycosides (amikacin, gentamicin and kanamycin) and benzalkonium chloride with fractional inhibitory concentration index values of 0.33–0.44 and 0.29 against *Staphylococcus aureus*. Moreover, low concentrations (0.5–1.3 mg/mL) of F105 restored the antimicrobial efficacy of gentamicin and ampicillin against *S. aureus* biofilms [43]. This indicated that application of furanones increased the susceptibility *Salmonella* to the antibiotics. In the present study, we tested the effects of floridoside and tetracycline against *S.* Enteritidis. Results showed that floridoside potentiated the activity of tetracycline against *S.* Enteritidis (Figure 3). Interestingly, in *Pseudomonas aeruginosa* quorum sensing has been shown to regulate efflux pumps (i.e., demonstrated mediators of antibiotic resistance). Accumulation of quorum sensing, auto-inducers (C4-HSL) in the medium has been shown to increase the transcription of the multi-drug-resistant-pump MexAB-OprM [44]. Relatedly, in the present study, floridoside might have inhibited quorum sensing in *Salmonella*, which could have repressed efflux-related gene expression. Interference with efflux activity would have resulted in the accumulation of tetracycline within the cell, eventually leading to cell death. Thus, this finding demonstrated that selected seaweed compounds can be used as effective alternatives, in combination, to increase both the useful life and reduce the rates of effective concentrations of over-used antibiotics. Despite the significant progress in several pathogen control strategies, the incidence of *Salmonella* Enteritidis in poultry and its subsequent transmission to the human food chain has continued to be a food safety issue [45]. More importantly, the worldwide emergence of several resistant strains of *Salmonella* emphasizes a major food safety hazard. In poultry, *Salmonella* is responsible for either clinical diseases or asymptomatic subclinical infections, the latter is referred to as "carriers" [46]. In earlier poultry studies, it has been shown that subclinical infection in chickens can be persistent for >22 weeks. Thus, carriers play a vital role in the perpetuation of *Salmonella* transmission in the livestock and environment, specifically by shedding the pathogen in their feces without demonstrating any clinical disease symptoms [47]. Other means of *Salmonella* transmission include vertical and horizontal transmission. Vertical transmission, which involves passage of pathogen from parents to progeny, is critical in poultry production especially related to *Salmonella* infections caused by *S.* Enteritidis. It has been demonstrated that Enteritidis has higher affinity to the reproductive system of the layer hens as compared to other serovar [48–50]. Though several antimicrobials can be used as therapeutics against such zoonotic pathogens, an increasingly high resistance towards such antimicrobials point out the need to find natural alternatives for use in animal feeds and supplements.

In *Esherichia coli* (*E. coli*), the quorum sensing regulator SdiA was shown to control multi-drug-resistance by functioning as a positive regulator of the multi-drug-resistance-pump AcrAB. Over-production of SdiA was shown to increase the levels of AcrAB leading to multi-drug-resistance [51]. Previously, we have shown that crude seaweed extracts down-regulated the expression of SdiA [34] and floridoside was previously reported as a quorum sensing inhibitor [39]. Therefore, the possible mode of action of floridoside could be the inhibition of bacterial quorum sensing resulting in increased susceptibility of *Salmonella* to the antibiotics. Moreover, as quorum-sensing inhibitors do not cause bacterial cell death, the selection pressure for development of resistance could be immensely reduced in the pathogenic bacteria.

Tetracycline inhibits protein synthesis in bacteria by binding to the 30S subunit of the ribosome. Bacteria can acquire tetracycline resistance by enzymatic inactivation of the drug or by increasing efflux-pump activity. Multi-drug-efflux-pumps are membrane proteins that utilize cellular energy to transport antibiotics from the cells to the external environment [52]. In the present study, the relative transcript level of efflux-related genes of *Salmonella* Enteritidis, namely *marA*, *arcB* and *ramA*, were significantly repressed by the combined treatment of floridoside and tetracycline, as compared to control antibiotics alone (Figure 4). Reduced expression of efflux-related genes indicated a decrease in the efficiency of *Salmonella* to efflux tetracycline from the cells [53]. Thus, in the presence of floridoside, the efflux of tetracycline would have been reduced, resulting in the accumulation of tetracycline to a level which could potentially inhibit protein synthesis in the cell, thus eventually leading to cell death.

5. Conclusions

To conclude, this research indicated that extracts and pure compounds from the cultivated red seaweeds *Chondrus crispus* and *Sarcodiotheca gaudichaudii* could be used to enhance the activity of antibiotics which are most commonly used in poultry production. The extracts and compounds can work in combination with the sub-lethal doses of tetracycline and streptomycin in order to potentiate their antimicrobial activity. The proposed mode of action for the combined effects was that floridoside could inhibit the quorum sensing of *Salmonella*, repressing the efflux-related gene expression, resulting in cellular accumulation of tetracycline, ultimately leading to bacterial cell death. Taken together, these findings showed that specific seaweed compounds can be used to increase the lifetime of existing antibiotics. Further research needs to be carried out to understand the structure-activity relationship of floridoside and tetracycline, which enhanced the antimicrobial activity against *Salmonella*. This will further help to determine specific targets of floridoside in *Salmonella* that result in cell death and verify the role of quorum sensing in the inhibitory activity of floridoside and tetracycline. Our previous studies identified the antimicrobial activity of red seaweeds *Chondrus crispus* and *Sarcodiotheca* in vitro and its successful translation into in vivo in chickens [34,54,55]. In layer hens, since red seaweed responses were comparable to antibiotic (aureomycin) action, some red seaweeds can be used either as an organic feed alternative to antibiotics or in combination with reduced rates of antibiotic inclusion. As the CC and SG red seaweeds potentiated the activity of antibiotics in-vitro, it would be worthwhile to try various combinations of antibiotics and these seaweeds at different dosages in chickens. A feed additive that could lower the effective required dose of antibiotics could be extremely useful in decreasing the consumption of antibiotics in commercial poultry farms and may assist with the reduced incidence of further bacterial resistance to antibiotics.

Supplementary Materials: The following are available online at http://www.mdpi.com/2077-1312/8/7/511/s1, Figure S1: ^{1}H NMR spectrum of SWE A) *Sarcodiotheca gaudichaudii* B) *Chondrus crispus*. ^{1}H NMR. signals correspond to F—floridoside, I—isethionic acid, T—taurine and C—L-Citrulline, Figure S2: ^{1}H NMR spectrum of floridoside isolated from *Chondrus crispus*.

Author Contributions: Conceptualization, B.P.; data curation, G.K.; formal analysis, G.K., B.R. and B.P.; funding acquisition, B.P.; investigation, G.K., B.R., A.C., A.H.B., J.H. and B.P.; methodology, G.K., A.C. and A.H.B.; project administration, B.P.; resources, J.H. and B.P.; supervision, A.C., B.R., G.S. and B.P.; writing—original draft, G.K.; writing—review and editing, A.C., B.R., G.S., A.H.B. and B.P. All authors have read and agreed to the published version of the manuscript.

Funding: This research was funded by Natural Science and Engineering Research Council of Canada-IPS II (445754).

Acknowledgments: This work was supported by Natural Science and Engineering Research Council of Canada-IPS II (445754), Ottawa, ON, Canada, Nova Scotia Department of Agriculture and Marketing, NS, Canada, and Acadian Seaplants Limited, Dartmouth, NS, Canada, grant to B.P.

Conflicts of Interest: The authors declare that the research was conducted in the absence of any commercial or financial relationships that could be construed as a potential conflict of interest.

References

1. Brenes, A.; Roura, E. Essential oils in poultry nutrition: Main effects and modes of action. *Anim. Feed Sci. Technol.* **2010**, *158*, 1–14. [CrossRef]
2. Venkitanarayanan, K.; Kollanoor-Johny, A.; Darre, M.J.; Donoghue, A.M.; Donoghue, D.J. Use of plant-derived antimicrobials for improving the safety of poultry products1. *Poult. Sci.* **2013**, *92*, 493–501. [CrossRef] [PubMed]
3. Abd El-Aze, N.A.; El-Daly, E.F.; Hassan, H.M.A.; Youssef, A.W.; Mohamed, M.A. Histological Response of Broilers Immune Related Organs to Feeding Different Direct Fed Microbials. *Int. J. Poult. Sci.* **2015**, *14*, 331–337. [CrossRef]

4. Centers for Disease Control and Prevention. Food and Food Animals | Antibiotic/Antimicrobial Resistance | CDC. Available online: https://www.cdc.gov/drugresistance/food.html (accessed on 10 May 2020).
5. Van Boeckel, T.P.; Gandra, S.; Ashok, A.; Caudron, Q.; Grenfell, B.T.; Levin, S.A.; Laxminarayan, R. Global antibiotic consumption 2000 to 2010: An analysis of national pharmaceutical sales data. *Lancet Infect. Dis.* **2014**, *14*, 742–750. [CrossRef]
6. Van Boeckel, T.P.; Brower, C.; Gilbert, M.; Grenfell, B.T.; Levin, S.A.; Robinson, T.P.; Teillant, A.; Laxminarayan, R. Global trends in antimicrobial use in food animals. *Proc. Natl. Acad. Sci. USA* **2015**, *112*, 5649–5654. [CrossRef]
7. Van Boeckel, T.P.; Pires, J.; Silvester, R.; Zhao, C.; Song, J.; Criscuolo, N.G.; Gilbert, M.; Bonhoeffer, S.; Laxminarayan, R. Global trends in antimicrobial resistance in animals in low- and middle-income countries. *Science* **2019**, *365*. [CrossRef]
8. U.S. Food and Drug Administration. FDA Annual Summary Report on Antimicrobials Sold or Distributed in 2012 for Use in Food-Producing Animals. Available online: http://www.fda.gov/AnimalVeterinary/NewsEvents/CVMUpdates/ucm416974.htm (accessed on 10 May 2020).
9. U.S. Food and Drug Administration. Poultry and Eggs: Background. Available online: http://www.ers.usda.gov/topics/animal-products/poultry-eggs/background.aspx (accessed on 10 May 2020).
10. Canadian Food Inspection Agency. Index of Medicating Ingredients Approved by Livestock Species—Canadian Food Inspection Agency. Available online: https://www.inspection.gc.ca/animal-health/livestock-feeds/medicating-ingredients/mib/livestock-species/eng/1522783196554/1522783196850 (accessed on 10 May 2020).
11. Collier, C.T.; Van der Klis, J.D.; Deplancke, B.; Anderson, D.B.; Gaskins, H.R. Effects of tylosin on bacterial mucolysis, *Clostridium perfringens* colonization, and intestinal barrier function in a chick model of necrotic enteritis. *Antimicrob. Agents Chemother.* **2003**, *47*, 3311–3317. [CrossRef]
12. Roura, E.; Homedes, J.; Klasing, K.C. Prevention of immunologic stress contributes to the growth-permitting ability of dietary antibiotics in chicks. *J. Nutr.* **1992**, *122*, 2383–2390. [CrossRef]
13. Starr, M.P.; Reynolds, D.M. Streptomycin resistance of coliform bacteria from turkeys fed streptomycin. *Am. J. Public Health Nations Health* **1951**, *41*, 1375–1380. [CrossRef]
14. Bates, J.; Jordens, J.Z.; Griffiths, D.T. Farm animals as a putative reservoir for vancomycin-resistant enterococcal infection in man. *J. Antimicrob. Chemother.* **1994**, *34*, 507–514. [CrossRef]
15. Castanon, J.I.R. History of the use of antibiotic as growth promoters in European poultry feeds. *Poult. Sci.* **2007**, *86*, 2466–2471. [CrossRef] [PubMed]
16. Lewis, M.; Inman, C.F.; Bailey, M. Review: Postnatal development of the mucosal immune system and consequences on health in adulthood. *Can. J. Anim. Sci.* **2010**, *90*, 129–136. [CrossRef]
17. Yan, G.L.; Guo, Y.M.; Yuan, J.M.; Liu, D.; Zhang, B.K. Sodium alginate oligosaccharides from brown algae inhibit *Salmonella* Enteritidis colonization in broiler chickens. *Poult. Sci.* **2011**, *90*, 1441–1448. [CrossRef] [PubMed]
18. Filho, R.A.C.P.; de Paiva, J.B.; Argüello, Y.M.S.; da Silva, M.D.; Gardin, Y.; Resende, F.; Junior, A.B.; Sesti, L. Efficacy of several vaccination programmes in commercial layer and broiler breeder hens against experimental challenge with *Salmonella enterica* serovar Enteritidis. *Avian Pathol.* **2009**, *38*, 367–375. [CrossRef]
19. Centers for Disease Control and Prevention (CDC). Biggest Threats and Data, Antibiotic Antimicrobial Resistance. Available online: https://www.cdc.gov/drugresistance/biggest-threats.html (accessed on 10 May 2020).
20. Amábile-Cuevas, C.F. Antibiotic resistance: From Darwin to Lederberg to Keynes. *Microb. Drug Resist.* **2013**, *19*, 73–87. [CrossRef]
21. Hawkey, P.M. Why are bacteria resistant to antibiotics? *Intensive Care Med.* **2000**, *26*, S009–S013.
22. Nikaido, H. Multidrug Resistance in Bacteria. *Annu. Rev. Biochem.* **2009**, *78*, 119–146. [CrossRef]
23. Nikaido, H. Prevention of drug access to bacterial targets: Permeability barriers and active efflux. *Science* **1994**, *264*, 382–388. [CrossRef]
24. Mah, T.F.; Pitts, B.; Pellock, B.; Walker, G.C.; Stewart, P.S.; O'Toole, G.A. A genetic basis for *Pseudomonas aeruginosa* biofilm antibiotic resistance. *Nature* **2003**, *426*, 306–310. [CrossRef]
25. Khan, S.; Tøndervik, A.; Sletta, H.; Klinkenberg, G.; Emanuel, C.; Onsøyen, E.; Myrvold, R.; Howe, R.A.; Walsh, T.R.; Hill, K.E.; et al. Overcoming drug resistance with alginate oligosaccharides able to potentiate the action of selected antibiotics. *Antimicrob. Agents Chemother.* **2012**, *56*, 5134–5141. [CrossRef]
26. Hussin, W.A.; El-Sayed, W.M. Synergic interactions between selected botanical extracts and tetracycline against gram positive and gram negative bacteria. *J. Biol. Sci.* **2011**, *11*, 433–441. [CrossRef]

27. Singh, A.P.; Prabha, V.; Rishi, P. Value Addition in the Efficacy of Conventional Antibiotics by Nisin against *Salmonella*. *PLoS ONE* **2013**, *8*, e76844. [CrossRef] [PubMed]

28. Hierowski, M. Inhibition of protein synthesis by chlortetracycline in the *E. coli* in In vitro system. *Proc. Natl. Acad. Sci. USA* **1965**, *53*, 594–599. [CrossRef] [PubMed]

29. Chovanová, R.; Mezovská, J.; Vaverková, Š.; Mikulášová, M. The inhibition the Tet(K) efflux pump of tetracycline resistant *Staphylococcus epidermidis* by essential oils from three Salvia species. *Lett. Appl. Microbiol.* **2015**, *61*, 58–62. [CrossRef]

30. Reed, R.H.; Collins, J.C.; Russell, G. Effects of Salinity upon Galactosyl-Glycerol Content and Concentration of the Marine Red Alga *Porphyra purpurea* (Roth) C.Ag. *J. Exp. Bot.* **1980**, *31*, 1539–1554. [CrossRef]

31. Pardoe, I.; Hartley, C. Carbohydrates such as Alpha-D-Galactopyranosyl-1, 2-glycerol and/or Isomers, Obtained by Solvent Extraction from Algae, for Use as Viricides or Anticarcinogenic Agents. U.S. Patent Application 10/344,812, 2001.

32. Clinical and Laboratory Standards Institute (CLSI). Methods for dilution antimicrobial susceptibility tests for bacteria that grow aerobically. In *Approved Standard*, 9th ed.; CLSI document M07-A9; CLSI: Wayne, PA, USA, 2013.

33. Jorgensen, J.; Ferraro, M. Antimicrobial susceptibility testing: General principles and contemporary practices. *J. Clin. Infect. Dis.* **2009**, *49*, 1749–1755. [CrossRef]

34. Kulshreshtha, G.; Borza, T.; Rathgeber, B.; Stratton, G.S.; Thomas, N.A.; Critchley, A.; Hafting, J.; Prithiviraj, B. Red seaweeds *Sarcodiotheca gaudichaudii* and *Chondrus crispus* down regulate virulence factors of *Salmonella* enteritidis and induce immune responses in *Caenorhabditis elegans*. *Front. Microbiol.* **2016**, *7*. [CrossRef]

35. Su, L.-H.; Chiu, C.-H.; Chu, C.; Ou, J.T. Antimicrobial Resistance in Nontyphoid *Salmonella* Serotypes: A Global Challenge. *Clin. Infect. Dis.* **2004**, *39*, 546–551. [CrossRef]

36. Guerra, B.; Junker, E.; Miko, A.; Helmuth, R.; Mendoza, M.C. Characterization and localization of drug resistance determinants in multidrug-resistant, integron-carrying *Salmonella* enterica serotype typhimurium strains. *Microb. Drug Resist.* **2004**, *10*, 83–91. [CrossRef]

37. Infection Disease Society of America (ISDA). Review Finds Antibiotic Development Increased, but Insufficient. Available online: https://www.idsociety.org/news--publications-new/articles/2019/review-finds-antibiotic-development-increased-but-insufficient/ (accessed on 10 May 2020).

38. Talbot, G.H.; Jezek, A.; Murray, B.E.; Jones, R.N.; Ebright, R.H.; Nau, G.J.; Rodvold, K.A.; Newland, J.G.; Boucher, H.W. The Infectious Diseases Society of America's 10 × '20 Initiative (10 New Systemic Antibacterial Agents US Food and Drug Administration Approved by 2020): Is 20 × '20 a Possibility? *Clin. Infect. Dis.* **2019**, *69*, 1–11. [CrossRef]

39. Kim, J.S.; Kim, Y.H.; Seo, Y.W.; Park, S. Quorum sensing inhibitors from the red alga, *Ahnfeltiopsis flabelliformis*. *Biotechnol. Bioprocess. Eng.* **2007**, *12*, 308–311. [CrossRef]

40. Liu, H.B.; Koh, K.P.; Kim, J.S.; Seo, Y.; Park, S. The effects of betonicine, floridoside, and isethionic acid from the red alga *Ahnfeltiopsis flabelliformis* on quorum-sensing activity. *Biotechnol. Bioprocess. Eng.* **2008**, *13*, 458–463. [CrossRef]

41. Janssens, J.C.A.; Steenackers, H.; Robijns, S.; Gellens, E.; Levin, J.; Zhao, H.; Hermans, K.; De Coster, D.; Verhoeven, T.L.; Marchal, K.; et al. Brominated furanones inhibit biofilm formation by *Salmonella* enterica serovar Typhimurium. *Appl. Environ. Microbiol.* **2008**, *74*, 6639–6648. [CrossRef] [PubMed]

42. Pan, J.; Bahar, A.A.; Syed, H.; Ren, D. Reverting antibiotic tolerance of *Pseudomonas aeruginosa* PAO1 Persister Cells by (Z)-4-bromo-5-(bromomethylene)-3-methylfuran-2(5H)-one. *PLoS ONE* **2012**, *7*, e45778. [CrossRef] [PubMed]

43. Sharafutdinov, I.S.; Trizna, E.Y.; Baidamshina, D.R.; Ryzhikova, M.N.; Sibgatullina, R.R.; Khabibrakhmanova, A.M.; Latypova, L.Z.; Kurbangalieva, A.R.; Rozhina, E.V.; Klinger-Strobel, M.; et al. Antimicrobial effects of sulfonyl derivative of 2(5H)-Furanone against planktonic and biofilm associated methicillin-resistant and -susceptible *Staphylococcus aureus*. *Front. Microbiol.* **2017**, *8*, 2246. [CrossRef]

44. Maseda, H.; Sawada, I.; Saito, K.; Uchiyama, H.; Nakae, T.; Nomura, N. Enhancement of the mexAB-oprM Efflux Pump Expression by a quorum-Sensing autoinducer and its cancellation by a regulator, MexT, of the mexEF-oprN efflux pump operon in *Pseudomonas aeruginosa*. *Antimicrob. Agents Chemother.* **2004**, *48*, 1320–1328. [CrossRef]

45. Rajić, A.; Waddell, L.A.; Sargeant, J.M.; Read, S.; Farber, J.; Firth, M.J.; Chambers, A. An overview of microbial food safety programs in beef, pork, and poultry from farm to processing in Canada. *J. Food Prot.* **2007**, *70*, 1286–1294. [CrossRef]

46. Zamora-Sanabria, R.; Alvarado, A.M. Preharvest *Salmonella* Risk Contamination and the Control Strategies. In *Current Topics in Salmonella and Salmonellosis*; InTechOpen: Rijeka, Croatia, 2017.

47. Jajere, S.M. A review of *Salmonella* enterica with particular focus on the pathogenicity and virulence factors, host specificity and adaptation and antimicrobial resistance including multidrug resistance. *Vet. World* **2019**, *12*, 504–521. [CrossRef]

48. Okamura, M.; Miyamoto, T.; Kamijima, Y.; Tani, H.; Sasai, K.; Baba, E. Differences in abilities to colonize reproductive organs and to contaminate eggs in intravaginally inoculated hens and in vitro adherences to vaginal explants between *Salmonella* enteritidis and other *Salmonella* serovars. *Avian Dis.* **2001**, *45*, 962–971. [CrossRef]

49. Foley, S.L.; Johnson, T.J.; Ricke, S.C.; Nayak, R.; Danzeisen, J. *Salmonella* Pathogenicity and Host Adaptation in Chicken-Associated Serovars. *Microbiol. Mol. Biol. Rev.* **2013**, *77*, 582–607. [CrossRef]

50. Ebers, K.L.; Zhang, C.Y.; Zhang, M.Z.; Bailey, R.H.; Zhang, S. Transcriptional profiling avian beta-defensins in chicken oviduct epithelial cells before and after infection with Salmonella enterica serovar Enteritidis. *BMC Microbiol.* **2009**, *9*, 153. [CrossRef]

51. Rahmati, S.; Yang, S.; Davidson, A.L.; Zechiedrich, E.L. Control of the AcrAB multidrug efflux pump by quorum-sensing regulator SdiA. *Mol. Microbiol.* **2002**, *43*, 677–685. [CrossRef] [PubMed]

52. Littlejohn, T.G.; Paulsen, I.T.; Gillespie, M.T.; Tennent, J.M.; Midgley, M.; Jones, I.G.; Purewal, A.S.; Skurray, R.A. Substrate specificity and energetics of antiseptic and disinfectant resistance in *Staphylococcus aureus*. *FEMS Microbiol. Lett.* **1992**, *95*, 259–265. [CrossRef]

53. Poole, K. Multidrug Efflux Pumps and Antimicrobial Resistance in *Pseudomonas aeruginosa* and Related Organisms. *J. Mol. Microbiol. Biotechnol.* **2001**, *3*, 255–263.

54. Kulshreshtha, G.; Rathgeber, B.; MacIsaac, J.; Boulianne, M.; Brigitte, L.; Stratton, G.; Thomas, N.A.; Critchley, A.T.; Hafting, J.; Prithiviraj, B. Feed supplementation with red seaweeds, *Chondrus crispus* and *Sarcodiotheca gaudichaudii*, reduce *Salmonella* Enteritidis in laying hens. *Front. Microbiol.* **2017**, *8*. [CrossRef] [PubMed]

55. Kulshreshtha, G.; Rathgeber, B.; Stratton, G.; Thomas, N.; Evans, F.; Critchley, A.; Hafting, J.; Prithiviraj, B. Immunology, health, and disease: Feed supplementation with red seaweeds, *Chondrus crispus* and *Sarcodiotheca gaudichaudii*, affects performance, egg quality, and gut microbiota of layer hens. *Poult. Sci.* **2014**, *93*. [CrossRef] [PubMed]

Journal of
*Marine Science
and Engineering*

Article

Effect of Marine Macroalga *Enteromorpha* sp. Enriched with Zn(II) and Cu(II) ions on the Digestibility, Meat Quality and Carcass Characteristics of Growing Pigs

Izabela Michalak [1],*, Katarzyna Chojnacka [1] and Daniel Korniewicz [2]

[1] Department of Advanced Material Technologies, Faculty of Chemistry, Wrocław University of Science and Technology, Smoluchowskiego 25, 50-372 Wrocław, Poland; katarzyna.chojnacka@pwr.edu.pl
[2] Cargill (Polska) Sp. z o.o., 2/4 Rolna Str., 62-280 Kiszkowo, Poland; daniel.korniewicz@lnb.pl
* Correspondence: izabela.michalak@pwr.edu.pl; Tel.: +48-713202434

Received: 8 April 2020; Accepted: 11 May 2020; Published: 13 May 2020

Abstract: In the present study, the effect of macroalga *Enteromorpha* sp. enriched with Zn(II) and Cu(II) ions on daily amounts of feces and urine excreted by growing pigs, apparent fecal nutrient digestibility and daily nitrogen balance and retention, meat quality and the slaughter value of carcasses was examined. The duration of feeding experiments was 87 days. In the control group, the requirement for zinc and copper was covered by inorganic salts, whereas in the experimental group algae enriched with these elements via biosorption were supplemented. No effect of *Enteromorpha* sp. on the increase in digestibility of dry matter, dry organic matter, crude protein, crude fat and nitrogen-free extractives was observed. Statistically significant differences concerned only the digestibility of crude ash. The daily amount of excreted feces and urine did not differ significantly between groups. Meat from pigs in the algal group was characterized by a lower water absorption and drip loss and contained less fat and more protein than meat from the control group. Furthermore, a slight darkening of the meat was observed. The weight of the liver was lower in pigs from the algal group. Enriched macroalga *Enteromorpha* sp. may be introduced into pig nutrition as a feed material as an alternative to inorganic salts.

Keywords: green macroalgae; microelements; feed additive; feeding experiment; growing pigs

---

1. Introduction

Seaweeds (called also macroalgae) have been used for millennia as a feed supplement in order to improve animal nutrition and productivity [1]. Macroalgae are recognized as a valuable raw material for the production of feed additives due to their enormous biodiversity, which can be exploited, and the fact that seaweeds are widely used as foods, fertilizers, components of pharmaceuticals, cosmetics, etc. [2,3]. Limited animal studies suggest that seaweeds may be also used in pig nutrition in order to ameliorate gut health, to boost the immune system and growth performance. In the present paper, the application of marine green macroalgae as a feed additive for pigs is proposed. In the review papers of Makkar et al. (2016) [4], Angell et al. (2016) [5], Corino et al. (2019) [6] and Øverland et al. (2019) [7], it was shown that seaweeds can serve as a source of active compounds for pigs, such as polysaccharides, proteins and amino acids (lysine, histidine, isoleucine, leucine, arginine, methionine, phenylalanine, threonine, tryptophan, valine, tyrosine, alanine, glutamine, asparagine), lipids including omega 3 and 6 fatty acids, vitamins (E, A, C, B_1, B_2, B_3), minerals (Ca, Mg, P, K, Na, Mn, Zn, Fe, Cu, I, Se, Co), phenolic compounds (e.g., phlorotannins) etc. [4–7]. These compounds demonstrate positive health effects, such as prebiotic, antibacterial, antioxidant, anti-inflammatory and immunostimulant effects. [4,6]. Seaweeds in pigs' nutrition show a beneficial influence on the digestibility of feed (nitrogen, polysaccharides, fiber, dry matter, organic matter), health and welfare of pigs [6]. Due to the presence of sulfated polysaccharides, such as alginates, ulvans and fucoidans, seaweeds also play prebiotic functions and positively modulate the intestinal microbiota [6,8]. Seaweeds can be also considered as potential antibiotic replacers in pigs [4,6]. Healthy and valuable feed is responsible for animal health and thus the high quality of animal products such as meat.

Seaweeds in pig feed are used in different forms—as a dried biomass, as extracted compounds (mainly polysaccharides—alginates, laminarin and fucoidan), or seaweed extracts [6]. Among brown (*Phaeophyceae*), red (*Rhodophyceae*) and green seaweeds (*Chlorophyceae*), brown algae dominate in pig nutrition (e.g., *Ascophyllum nodosum* [4,9,10], *Fucus vesiculosus* [4], *Laminaria japonica* [11], *Laminaria* sp., as well as laminarin and fucoidan extracted from *Laminaria* [4,8]). In this work, we propose to utilize the valuable composition and properties of marine green macroalgae (*Enteromorpha* sp.) for the production of feed additives with microelements. Dry seaweeds are able to bind efficiently metal ions from the aqueous solutions to the functional groups that are present on the surface of biomass [12–14]. In the present study, we enriched the algal biomass with microelements—Cu(II) and Zn(II)—using a rapid and reversible process called biosorption [13,14]. These two elements were chosen, since they are crucial for animals and are known to exert positive influence on growth performance of young pigs [15]. On the other hand, there are some concerns associated with the increase in Cu and Zn load in the environment, derived mainly from piggery effluents, which could have adverse effects on the soil microbiota and potentially cause a decline in soil fertility and pasture, as well as on crop yields [16]. Therefore, there is a need to reduce the level of copper and zinc in the diet of growing pigs without detrimental effects on the production and mineral status. The solution is to replace traditionally used inorganic salts, which are not easily absorbed by organic form of minerals that can increase the mineral absorption and retention in pigs [17]. In our previous work, we evaluated the effect of the enriched with Zn(II) and Cu(II) ions macroalga *Enteromorpha* sp. on the mineral composition of blood, meat, liver, feces and urine of growing pigs, production parameters, as well as biochemical markers such as crude protein, albumins, glucose, urea, liver enzyme: aspartate aminotransferase, alanine aminotransferase, gamma-glutamyl transpeptidase, total cholesterol and its fractions: high density lipoprotein (HDL), low density lipoprotein (LDL) and triglycerides. It was found that the bioavailability of microelements to pigs from algae was higher than from inorganic salts that were supplemented in the control group [18]. Moreover, Dierick et al. (2009) suggested that brown seaweed—*Ascophyllum nodosum*—may be introduced into pigs' diets as a feed material with a double role: the improvement of pig gut health and performance, and the iodine enrichment of porcine tissues [9].

The aim of the present study was to examine the effect of enriched with Cu(II) and Zn(II) ions *Enteromorpha* sp. on the daily amount of feces and urine excreted by growing pigs, apparent fecal nutrient digestibility, daily nitrogen balance and retention, meat quality and slaughter value of carcasses.

2. Materials and Methods

2.1. Raw Material

The alga *Enteromorpha* sp. was collected from the Baltic Sea (Niechorze—Poland) and identified in the Department of Botany and Plant Ecology of Wrocław University of Environmental and Life Sciences (Poland). This macroalga dominates in the macrophytobenthos on the Polish coast. Large quantities of seaweeds result from eutrophication. The touristic attractiveness of the seaside resorts nearby is therefore reduced [19]. On the other hand, this edible macroalga is characterized by a high nutritional value. It has been shown that *Enteromorpha* sp. is rich in lipids (in % of dry mass (DM)): from 3.47 ± 1.76 to 4.36 ± 2.17 [20], in proteins: from 9.42 ± 4.62 to 20.6 ± 5.0 [21], in carbohydrates: from 29.1 ± 6.44 to 39.8 ± 11.2 [21] and minerals [18,22]. The examined *Enteromorpha* sp. from the Baltic Sea contains micro- and macroelements in the amounts: Co 1.18 ± 0.18 mg/kg, Cu 2.17 ± 0.33 mg/kg, Fe 705 ± 106 mg/kg,

Mn 51.0 ± 7.6 mg/kg, Zn 15.2 ± 2.3 mg/kg, Ca 9040 ± 1810 mg/kg, K 15,400 ± 3100 mg/kg, Mg 20,500 ± 4100 mg/kg, Na 19,400 ± 3900 mg/kg [18].

2.2. Production of Algal Feed Additives

Baltic macroalga was enriched with Cu(II) and Zn(II) ions through biosorption. The solutions of microelements were prepared in 40 L of tap water (by dissolving appropriate amounts of $CuSO_4 \cdot 5H_2O$ and $ZnSO_4 \cdot 7H_2O$ (Avantor Performance Materials Poland S.A., Gliwice, Poland). Biosorption was carried out at room temperature for 4 h and the pH of solutions with a concentration of Cu(II) and Zn(II) equal to 300 mg/L was 5. The content of dry biomass was 1 g/L [18]. The best process parameters were established in our previous research [12]. After biosorption, the biomass of macroalgae was dried in air and then crushed in a blender. The content of Cu and Zn in the enriched algal biomass was determined using Inductively Coupled Plasma Optical Emission Spectrometry (ICP-OES) and was equal to 51.6 g of Cu/kg of dry mass (DM) and 56.4 g Zn/kg, respectively [18].

Three types of complete mixtures for growing pigs—"Starter"(S), "Grower" (G) and "Finisher" (F)—were produced and added to the standard feed, composed of ground wheat (S: 35%, G: 40% and F: 40%), ground barley (S: 41.7%, G: 43.4% and F: 47.9%), soybean meal (S: 15.5%. G: 11.5%, F: 8%), canola oil (S: 3.3%, G: 1.8%, F: 1.4%) and acidifier Lonacid Max (1017) (S: 0.5%, G: 0.3%, F: 0.2%), supplementary feed (S: 4.0%, G: 3.0%, F: 2.5%) [18,23]. The produced enriched algal biomass was sent to the company LNB Poland Ltd. (Poland), which was responsible for the preparation of premix—a source of trace elements and vitamins for growing pigs. The detailed chemical composition of the feed mixture is presented in Table 1.

Taking into account the content of Cu and Zn in the standard feed, the coverage of the requirement for these elements according to the Feeding Standards for Poultry and Swine (2005) [24] was calculated. The difference in the standard feed was supplemented by enriched algae (premix)—Table 2. The limiting factor was copper—the content of this microelement in the feed for growing pigs should not exceed 25 mg/kg feed.

The mineral composition of the feed used in the feeding experiments on growing pigs, both for the control group—microelements supplemented by inorganic salts (C), as well as for the experimental groups—microelements supplemented by enriched macroalga (MA) is presented in Table 3 [18].

2.3. Feeding Experiments on Growing Pigs

The feeding experiments on growing pigs were approved by the Second Local Ethical Committee on Animal Testing at Wrocław University of Environmental and Life Sciences. This work was carried out in accordance with EU Directive 2010/63/EU for animal experiments. These experiments were conducted at the Experimental Station of the Poznań University of Life Sciences in Gorzyń (Poland) and lasted for 87 days. The general scheme of these studies is shown in Figure 1.

The buildings where pigs were housed were cleaned and disinfected before experiments. The temperature inside the buildings was 16–18 °C. Natural and artificial lighting illuminated the whole area. Growing pigs originated from the following breeds: sow—the Polish Landrace/Polish Large White cross and boar—Hampshire/Pietrain cross. The study was conducted on two groups—the control and experimental. There were 12 piglets (eight barrows and four gilts) in each group. Both groups were fed with feed mixtures, which were characterized by the same content of nutrients, but in a different form. The control group received Zn and Cu as inorganic salts—$CuSO_4 \cdot 5H_2O$ and $ZnSO_4 \cdot 7H_2O$—and the experimental group *Enteromorpha* sp. enriched with Cu and Zn (from the basic mixture, inorganic forms of Zn and Cu were removed and replaced by enriched algae). Before the start of the study, each pig was marked by ear tagging and also dewormed (Dectomax®or Ivomec®). Animals were kept in individual pens in order to control the feed mixture intake [18,23].

Table 1. The chemical composition of feed mixtures for growing pigs (Reproduced with permission from Saeid et al., J. Appl. Phycol.; published by Springer, 2013 [23]).

Ingredient (in 1 kg of Mixture)	Unit	Type of Mixture		
		"Starter"	"Grower"	"Finisher"
Net energy	kcal	2340	2280	2281
Metabolizable energy	MJ	13.6	13.2	13.2
Dry mass	%	87.3	87.2	87.1
Crude protein	%	17.4	15.7	14.5
Crude fiber	%	3.00	2.80	3.50
Crude fat	%	5.00	3.10	3.20
Crude ash	%	5.10	4.30	3.70
N-free extractives	%	56.8	61.3	62.2
L-Lysine	%	1.17	0.93	0.85
Methionine	%	0.39	0.29	0.26
Methionine+Cysteine	%	0.71	0.60	0.55
L-Threonine	%	0.75	0.59	0.54
Tryptophan	%	0.23	0.20	0.16
Isoleucine	%	0.66	0.59	0.51
Calcium (Ca) total	%	0.73	0.68	0.60
Phosphorus (P) total	%	0.55	0.50	0.43
Mineral phosphorus (P)	%	0.16	0.15	0.13
Digestible phosphorus (P)	%	0.34	0.30	0.25
Phytase	FTU[a]	500	510	425
Sodium (Na)	%	0.20	0.20	0.14
Iron (Fe)[b]	mg	198	183	172
Manganese (Mn)[b]	mg	91	82	73
Copper (Cu)[b]	mg	167	25	22
Zinc (Zn)[b]	mg	157	148	126
Iodine (I)[b]	mg	1.66	1.49	1.26
Cobalt (Co)[b]	mg	0.88	0.81	0.68
Selenium (Se)[b]	mg	0.49	0.48	0.44
Vitamin A[c]	I.U.	16,000	12,000	10,000
Vitamin D_3[c]	I.U.	2000	1998	1665
Vitamin E[c]	mg	150	124	104
Vitamin K_3[c]	mg	4.00	1.80	1.50
Vitamin B_1[c]	mg	2.40	1.80	1.50
Vitamin B_2[c]	mg	6.40	4.80	4.00
Vitamin B_3 (Niacin)[c]	mg	32.0	24.0	20.0
Vitamin B_5 (Pantothenic acid)[c]	mg	16.0	12.0	10.0
Vitamin B_6[c]	mg	4.8	3.6	3.0
Vitamin B_{12}[c]	mcg	40.0	30.0	25.0
Vitamin C[c]	mg	100	100	83.3
Biotin[c]	mcg	160	120	100
Folic acid [c]	mg	3.20	2.40	2.00
Choline[c]	mg	350	250	208

[a] One "Phytase Unit" (FTU) is defined as that quantity of enzyme that will liberate inorganic phosphate at one micromole per minute from sodium phytate based on a 30-minute hydrolysis of sodium phytate at 37 °C and pH 5.5; [b] Microelements supplemented: Fe as $FeSO_4 \cdot H_2O$ 30%; Mn as MnO_2 60%; Cu as $CuSO_4 \cdot 5H_2O$ 25%; Zn as $ZnSO_4 \cdot H_2O$ 35%; I as Ca $(IO_3)_2 \cdot H_2O$ 62%, Co as $CoCO_3$ 21%; Se as Na_2SeO_3 5%; [c] Vitamins supplemented: retinyl acetate (A), cholecalciferol (D_3), DL α-tocopherol acetate (E), menadione sodium bisulfite (K), thiamine mononitrate (B_1), riboflavin (B_2), nicotinic acid; niacin (B_3), pantothenic acid; D-calcium pantothenate (B_5), pyridoxine hydrochloride (B_6), cyanocobalamin(B_{12}), ascorbic acid (C), D-biotin (biotin), folic acid, choline chloride (choline).

Table 2. Enriched with microelements *Enteromorpha* sp. added to the mixtures in order to cover the requirement for Cu and Zn (%).

Microelement	Requirement	"Starter"	"Grower"	"Finisher"
Cu	Requirement for Cu in the standard feed that should be covered by the feed additive	25%	84%	84%
	The coverage of the requirement by enriched algae (as a premix)	26%	100%	100%
Zn	Requirement for Cu in the standard feed that should be covered by the feed additive	28%	15%	15%
	The coverage of the requirement by enriched algae (as a premix)	32%	18%	18%

Table 3. The mineral composition of the feed for pigs in the control and experimental groups (Data from Michalak et al., Open Chem., De Gruyter Open, 2015 [18]).

Element	"Starter"		"Grower"		"Finisher"	
	C	MA	C	MA	C	MA
	mean ± SD (mg/kg DM)					
Ca	4412 ± 1484	4476 ± 698	4038 ± 787	3819 ± 581	3611 ± 595	2872 ± 732
Cu	13.3 ± 7.6	4.08 ± 0.62	9.06 ± 2.16	5.54 ± 1.06	11.3 ± 3.7	6.16 ± 4.9
Fe	206 ± 141	166 ± 87	149 ± 79	193 ± 56	205 ± 9	91.2 ± 63
K	3550 ± 716	3216 ± 444	3459 ± 593	3461 ± 566	3721 ± 270	2854 ± 576
Mg	788 ± 156	728 ± 107	767 ± 125	750 ± 118	846 ± 72	674 ± 123
Mn	82.0 ± 29.7	61.3 ± 16.3	66.4 ± 21.0	77.0 ± 15.5	63.7 ± 12.8	52.0 ± 16.7
Na	999 ± 187	909 ± 111	866 ± 118	890 ± 136	841 ± 84	745 ± 222
Zn	91.4 ± 22.9	73.0 ± 9.8	78.1 ± 17.5	76.0 ± 14.3	86.5 ± 11.4	62.3 ± 14.8

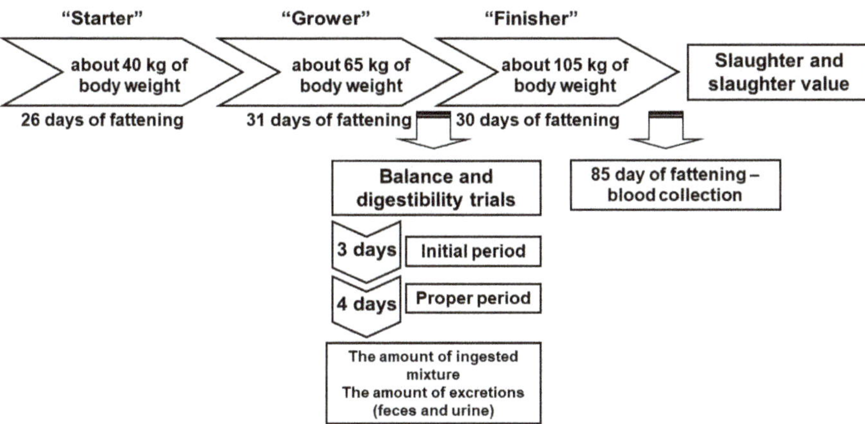

Figure 1. General scheme of feeding experiment on growing pigs.

During the production test, 12 barrows (six in each group) with a body weight of approximately 50–60 kg, were tested in balance and digestibility trials. Exactly 31 days after of feeding with "Grower" mixture, barrows were placed in individual metabolism cages, where they were fed with the same mixtures as during the whole experiment. The amount of the feed mixture was the same for all animals and it was ca. 2.5 kg per day. Every morning, the uneaten remains of the mixture were determined by weighing. The period of the first 3 days was treated as preliminary—the preparatory period after changing the living conditions. Over a further 4 days (proper period), the amount of ingested mixture and the quantity of feces and urine was recorded each day. Pigs urine flew into a special plastic tanks placed under the cages. Into these tanks, every day, 10 mL of 10% sulfuric acid was poured in order to bind the ammoniacal nitrogen. Pigs' feces were stopped on a grid placed under the grill of the pens. The daily collections of 10% pig feces and urine were collected in special jars with ground glass stoppers (urine) and in plastic bags (feces). The collected samples were stored in a refrigerator at 3–4 °C. Urine and feces collected during the period were thoroughly mixed.

At the end of fattening (after about 105 kg of body weight), from each group, 10 pigs (seven barrows and three gilts) were selected for slaughter, according to the standards in the meat industry—Minister of Agriculture and Rural Development dated April 2, 2004 (Polish Journal of Laws 2004.70.643). The slaughter procedure was carried out in a slaughterhouse by persons entitled to professional slaughter and by using acceptable methods of slaughter and killing of animals. The approved procedure involved the use of electronarcosis and the exsanguination of the pigs [23]. The post slaughter evaluation was also performed, which concerned hot carcass weight, carcass yield, slaughter yield, loin eye area, backfat thickness and the weight of the liver.

2.4. Analytical Methods

Nitrogen was determined by Kjeldahl's method, according to PN-EN ISO 5983–1:2006/AC:2009 (animal feeding stuffs—determination of nitrogen content and calculation of crude protein content—part 1: Kjeldahl method). In the wet fecal samples, the content of dry matter and nitrogen was determined. In dried fecal samples, crude fat, crude fiber, crude ash was measured and in the urine samples, the concentration of nitrogen was measured.

The content of meat in the carcass was measured with the use of an optical needle device—CGM apparatus (France). The measurement of pH was performed 1 and 24 hours after slaughter with the use of a Radiometer Copenhagen PHM80 Portable pH Meter with combined electrode. Electrical conductivity in the muscles after 24 h was determined by conductometer MP–03. The color of the meat was measured by Minolta Chroma Mater CR 300 (Konica Minolta Sensing, Inc., Japan) to detect the $L*a*b*$ values (where: L—color lightness, a—color value red, b—color value yellow).

The content of fat, water and protein in meat samples (*longissimus dorsi* muscle) was determined according to standard chemical methods [25]. The drip loss was calculated from the difference between the initial and final mass of the sample, and was placed in foil sack at a temperature of 4 °C for 48 h. IM-03 Pig Carcass Grading Apparatus was used to analyze the physical parameters of the meat such as the area of the loin eye and backfat thickness.

2.5. Statistical Methods

In our paper, two independent groups of pigs were compared: the first fed with inorganic salts as feed additive (the control group) and the second fed with enriched macroalgae (the experimental group). At first, we checked whether our dataset was well modeled by a normal distribution or not with the use of Shapiro–Wilk normality test. If the distribution of the dependent variables was non-normal, then the non-parametric Mann–Whitney U test was used. If the distribution of the dependent variables was normal, then for the two compared groups (of which the size was smaller than 30) the homogeneity of variance was checked with the use of the Brown–Forsyth test. If the variances were homogeneous, then the t test was chosen, if not, we used the Cochran–Cox test. The results were elaborated statistically by *Statistica* ver. 9.0. Results were considered significantly different when $P < 0.05$.

3. Results

The animals remained healthy throughout the experiment, as shown in our previous work, in which we examined the effects of enriched macroalgae on the growth performance of growing pigs—average feed intake, average weight gain and feed conversion ratio, as well as biochemical markers in the serum of examined pigs and mineral composition of blood, meat, liver, feces and urine of growing pigs. All the examined parameters were comparable in both examined groups and were not statistically significant [18].

3.1. Balance and Digestibility Trials

Balance and digestibility trials were conducted only in male individuals, which allowed for the separate collection of feces and urine. These trials usually start when the body weight reaches about 65 kg, which corresponds to the finishing feeding with "Grower" mixture. In this period, the best indicators, which concern protein, fat and mineral balance, are usually achieved. After this period of fattening, more intensive protein deposition and an increase in fat storage is observed. During the balance and digestibility trials, a "Grower" mixture in the control and experimental group contained the following components: dry weight (872 g/kg feed), total protein (157 g/kg), crude fat (31 g/kg), crude fiber (28 g/kg), crude ash (C—43 g/kg, MA—45 g/kg), nitrogen-free extractives (613 g/kg), minerals: Ca (68 g/kg), P total (50 g/kg), P digestible (30 g/kg), Cu (25 mg/kg), Zn (148 mg/kg), Mn (82 mg/kg), Fe (183 mg/kg), I (1.49 mg/kg), Co (0.81 mg/kg) and Se (0.48 mg/kg) [18].

3.2. Daily Amounts of Feces and Urine Excreted by Growing Pigs

Table 4 summarizes the results concerning the daily amount of feces and urine excreted by pigs during the collection period, which lasted 4 days. There were no statistically significant differences between the control and algal group, but the experimental group excreted 13% more feces and 20% less urine when compared with the control group. The nitrogen concentration in urine was 34% higher in the experimental than in the control group. Other differences were lower than 5%. These results are in agreement with data presented by Saeid et al. (2013) in analogous experiments carried out with microalga *Spirulina maxima* enriched with Cu(II), Zn(II) and Fe(II) ions as a feed additive [23].

Table 4. Daily amounts of feces and urine excreted by growing pigs in the collection period (4 days).

Specification	C	MA	p Value	Statistical Test
	Mean ± SD			
Feces				
Feces excreted (g)	765 ± 74	865 ± 254	0.749	Mann-Whitney
Dry matter (%)	32.7 ± 3.0	31.8 ± 5.9	0.750	Test t
Excreted dry matter (g)	249 ± 11	264 ± 29	0.251	Test t
Urine				
Urine excreted (g)	5 095 ± 1 035	4 079 ± 1 276	0.161	Test t
N (%)	0.427 ± 0.116	0.572 ± 0.154	0.0956	Test t
N excreted in urine (g)	20.8 ± 1.9	21.8 ± 2.7	0.483	Test t

3.3. Apparent Fecal Nutrient Digestibility (%) and Daily Nitrogen Balance and Retention

Table 5 presents the apparent fecal nutrient digestibility (%) and daily nitrogen balance and retention. No effect of enriched macroalgae on the increase in digestibility of dry matter, dry organic matter, crude protein, crude fat and nitrogen-free extractives was observed. A statistically significant difference concerned the digestibility of crude ash, which, in the experimental group, was 15% lower than in the control group. In the case of nitrogen retention and retention in relation to N intake (%), these parameters were lower by 5% in the algal group compared to the control.

3.4. Meat Quality and Slaughter Value of Carcass

In Table 6, the results, which concern the meat quality and slaughter value of carcasses are presented. A statistically significant difference was observed only for the liver weight, which, in the experimental group, was 14.5% lighter than in the control group. Beside this, groups of pigs were not significantly different in the characteristics of meat quality and slaughter value of their carcasses.

Table 5. Apparent fecal nutrient digestibility (%) and daily nitrogen balance and nitrogen retention.

Specification	C	MA	*p* Value	Statistical Test
	Mean ± SD			
Apparent Fecal Nutrient Digestibility (%)				
Dry matter	86.2 ± 1.3	84.9 ± 1.8	0.184	Test *t*
Dry organic matter	88.1 ± 1.2	87.0 ± 1.6	0.208	Test *t*
Total protein	87.4 ± 1.8	86.9 ± 3.7	0.689	Mann-Whitney
Total fat	77.0 ± 3.4	78.9 ± 8.6	0.630	Test *t*
Crude fiber	22.0 ± 8.2	18.8 ± 5.3	0.443	Test *t*
Crude ash	50.2 ± 4.4	42.5 ± 6.3	*0.0330*	Test *t*
Nitrogen-free extractives	91.9 ± 0.8	90.8 ± 1.1	0.0731	Test *t*
Daily Nitrogen Balance and Nitrogen Retention				
Nitrogen taken in the feed (g)	50.2 ± 0.0	50.2 ± 0.0	-	-
Nitrogen excreted (g) in:				
Feces	6.32 ± 0.90	6.53 ± 1.86	0.689	Mann-Whitney
Urine	20.8 ± 1.9	21.8 ± 2.7	0.486	Test *t*
Nitrogen retention (g)	23.1 ± 2.0	21.9 ± 2.0	0.331	Test *t*
Retention in relation to N intake (%)—absorption	45.9 ± 3.9	43.5 ± 3.9	0.311	Test *t*

Statistically significant differences ($p < 0.05$) were written in Italics.

Table 6. Assessment of slaughter value of carcass and meat quality.

Specification	C	MA	*p* Value	Statistical Test
	Mean ± SD			
Assessment of Slaughter Value of Carcass				
Hot carcass weight (kg)	90.7 ± 2.8	88.3 ± 4.1	0.143	Test *t*
Carcass yield (%)	54.7 ± 2.3	53.6 ± 3.5	0.408	Cochran-Cox
Loin eye area (cm²)	38.9 ± 4.8	37.5 ± 5.0	0.548	Test *t*
Weight of liver (g)	1 724 ± 227	1 474 ± 200	*0.0284*	Mann-Whitney
Average backfat thickness (mm)				
Over the shoulder	38.7 ± 6.7	37.7 ± 5.6	0.721	Test *t*
On the midback	20.2 ± 5.1	21.9 ± 5.7	0.489	Test *t*
On the rump I	20.2 ± 3.1	21.0 ± 5.3	0.650	Mann-Whitney
On the rump II	13.8 ± 2.6	14.3 ± 4.4	0.762	Test *t*
On the rump III	16.2 ± 3.6	16.2 ± 4.7	1.000	Test *t*
Assessment of meat quality				
pH $_1$ (after 45 minutes)	6.28 ± 0.24	6.26 ± 0.23	0.854	Test *t*
pH $_{24}$ (after 24 hours)	5.51 ± 0.0944	5.50 ± 0.0860	0.733	Test *t*
Water absorption (%)	32.9 ± 0.9	30.7 ± 3.4	0.0686	Cochran-Cox
Drip loss (%)	5.46 ± 2.13	5.05 ± 2.43	0.694	Test *t*
Marbling (degrees)	1.75 ± 0.26	1.70 ± 0.42	1.000	Mann-Whitney
Electrical conductivity (mS/cm²)	3.97 ± 1.15	4.04 ± 1.08	0.890	Test *t*
The content in muscles (%)				
Water	72.4 ± 1.1	72.5 ± 0.9	0.893	Test *t*
Fat	3.24 ± 0.89	2.62 ± 0.65	0.0821	Mann-Whitney
Protein	23.3 ± 0.7	23.8 ± 0.7	0.106	Test *t*
Color				
L (color lightness)	50.8 ± 1.7	51.1 ± 3.0	0.838	Test *t*
a (color value-red)	4.49 ± 0.72	4.52 ± 0.82	0.932	Test *t*
b (color value-yellow)	0.265 ± 0.806	0.0340 ± 1.38	0.653	Test *t*

Statistically significant differences ($p < 0.05$) are written in italics.

4. Discussion

In the present study, green seaweeds—*Enteromorpha* sp.—were examined in terms of their potential application in pig feed. Several parameters such as the daily amount of feces and urine excreted, apparent fecal nutrient digestibility, daily nitrogen balance and retention, meat quality and the slaughter value of carcasses were evaluated. It was shown that the digestibility of crude ash in the experimental group was 15% lower than in the control group. This difference was statistically significant. This may result from the naturally high level of this component in the biomass of *Enteromorpha* sp. collected from the Baltic Sea, which ranges from 19% to 32% [21], while the ash content in the dry matter in cereals ranges from 1.4% in maize to 2.7% in oats [26]. Moreover, crude fiber digestibility was 14.5% lower. Similar results (the same trend) were obtained by Saeid et al. (2013), who used microalga *Spirulina maxima* enriched with Cu(II), Zn(II) and Fe(II) ions as a feed additive for growing pigs [23]. Moreover, Dierick et al. (2009) found that the overall digestibility of nutrients in the pigs' diet seemed not to be negatively affected by seaweed supplementation (10 and 20 g/kg of feed). However, the apparent fecal nutrient digestibility of dry matter, dry organic matter, crude protein and crude ash in the experimental group was slightly higher than in the control group (for seaweed content 20 g/kg of feed it was: 3.2%, 3.5%, 14.7% and 21%, respectively). Only the digestibility of crude fat was 8.4% lower in the algal group than in the control group [9]. In the paper of Lynch et al. (2010), the digestibility coefficient of dry matter and dry organic matter in the control group and in all experimental groups (the content of *Laminaria hyperborea* extract was 0.7 g/kg; 1.4 g/kg; 2.8 g/kg and 5.6 g/kg) was the same and equal 89% and 91%, respectively [8].

We have also found that nitrogen retention and retention in relation to N intake (%) was lower in the algal group when compared to the control. Increased nitrogen excretion in the urine (control group—20.8 g, experimental group—21.8 g), required much more metabolic effort on the part of the pigs, as was seen in slightly weaker daily gains (by about 4% lower in algal than in control group) and in the greater feed conversion per kg of gain (about 2% lower in algal than in control group) [18]. In the control group, pigs excreted slightly less nitrogen (6.32 g) than in the experimental group (6.53 g). This can result in improved protein digestibility in the control group, which was 87.4%, while in the experimental group the protein digestibility was 86.9%. Moreover, Gardiner et al. (2008) observed that the nitrogen retention was 31% lower in the experimental group (addition of 2.5 g of *Ascophyllum nodosum* extract) than in the control group [10]. In our study, in the urine from both groups, significantly higher amounts of nitrogen were excreted when compared with feces. These results were also confirmed in the paper of Gardiner et al. (2008) who noted that the nitrogen content in urine in the control group was 29 g per day and in the experimental group 37 g per day (addition of 2.5 g of *Ascophyllum nodosum* extract), and in feces 8.7 g per day and 9.4 g per day, respectively [10]. Moreover, Lynch et al. (2010) examined the effect of dietary *Laminaria*-derived laminarin and fucoidan on nitrogen utilization in pigs. There was a quadratic response to seaweed extract on urinary nitrogen excretion, total nitrogen excretion and nitrogen retention [8].

Since meat and meat products are considered vital components of a healthy diet, increased consumer demand for food with reduced fat levels and cholesterol and an enhanced fatty acid profile is observed [27]. In the present work, it was shown that the meat quality and slaughter value of the carcass was comparable in both tested groups—control and enriched macroalgae. The only difference concerned the liver weight, which in the experimental group was 14.5% lighter than the control group. These results also confirm the values of liver enzymes (aspartate aminotransferase (AST), gamma-glutamyl transpeptidase (GGT)), which, in the experimental group, had lower values than the control group [18]. Saeid et al. (2013) also noted 5% lighter liver in the experimental group, in which pigs were fed with *Spirulina maxima* enriched with Cu(II), Zn(II) and Fe(II), when compared to the control group—microelements supplemented as inorganic salts [23]. Moreover, Svoboda et al. (2009) did not observe a difference in the hot weight of carcass in the group of pigs, which were fed with the addition of inorganic sodium selenite (78.6 ± 7.1 kg), and in the group fed with organic Se from Se-enriched alga—*Chlorella* spp. (78.5 ± 5.0 kg) [28]. The same applied in the case of two other microalgae, *Chlorella* and *Scenedesmus*, which were compared with pigs' diets containing fish meal. Carcass characteristics was comparable and the digestibility studies indicated that algae were low in digestible energy, but their protein was 70% digestible [29].

In our study, there was no statistically significant effect of algal additive on the average value of pH_1, pH_{24} (acidity of meat), water absorption, drip loss (respectively lower by 7.0% and 7.5% in the experimental group than in the control group—the smaller the leakage, the better). A decrease in water absorption by 5% and drip loss by 34 % was also noted by Saeid et al. (2013) in the case of the application of *Spirulina maxima* in the experimental group of pigs [23]. In the research of Suzuki et al. (2002), the influence of dried seaweed on meat production and its quality was examined. The cooking loss in the first group—dried seaweeds and breadcrumbs mixed with feed at a rate of 0.3% and 5%—and the second group—dried seaweed as an additive—was significantly lower than in the third group—breadcrumbs used as an additive—and the fourth group, where neither additive was used. The obtained results showed that the addition of dried seaweeds improved the meat quality [30]. Our results also confirmed data obtained by Svoboda et al. (2009) [28]. The pH of the meat (24 h after slaughter) was 5.66 in the group with inorganic Se and 5.68 in the group with Se-enriched alga. However, the leakage in the experimental group was 12.5% higher than in the control group [28].

The average values of carcass yield of the growing pigs in both groups did not differ from the average carcass yield of pigs in the national population, which in 2011 was 55.4% [31]. This indicates that they meet the current standards for carcass yield and can be used in the meat industry. However, the carcass yield of pigs in the control group (55%) was slightly higher than in the experimental group (54%). The reason may be greater nitrogen excretion from the body (both in feces and urine) in the experimental group. The control group was characterized by a better utilization of nitrogen, which was used to build muscle tissue, hence the higher carcass yield. Moreover, Sardi et al. (2006) found that the lean meat (%) in the control group of pigs (a maize/soybean diet) was slightly higher (49.1%) than in the experimental groups: the first (macroalgae added at 2.5 g/kg over the last 8 weeks prior to slaughtering) was 48.1%, the second (5 g/kg over the last 4 weeks prior to slaughtering) was 48.6%, and the third (2.5 g/kg over the last 4 weeks prior to slaughtering) was 48.7% [32].

Loin eye area is also related to the carcass yield, and in the experimental group it was lower than in the control group. The content of fat in muscle was by 19% lower in the experimental group than in the control group, which may be approved by consumers who are looking for products low in fat. The reduced content of fat in pigs from the algal group is associated with marbling (increased fat tissue), which was also lower in the experimental group. In the group of growing pigs fed with algae, a positive change in terms of reducing fat content and the growth of the desired protein was observed. Moderate marbling (the amount and distribution of intramuscular fat in muscle cross section) that is uniformly distributed is a desirable property [33]. The backfat thickness on the midback of growing pigs from the algal group was 8.4% higher than in pigs from the control group. There is a tendency for a decrease in thickness towards the rear of pigs' bodies. The average backfat thickness from 5 measurements (over the shoulder, on the midback and on the rump I, II, III) was greater in the experimental group (22.2 mm) than in the control group (21.8 mm). Other differences were lower than 5%. Moreover, a slight darkening of meat from pigs in the experimental group was observed. Choi et al. (2012) also reported an improvement in the color and sensory characteristics of reduced-fat pork patties as a result of the supplementation of pig feed with *Laminaria japonica* powder extract [11]. Nowadays, the meat color is one of the most important properties taken into account by consumers [34].

5. Conclusions

In the present study, the effect of macroalga *Enteromorpha* sp. enriched with Zn(II) and Cu(II) ions via biosorption on the daily amounts of feces and urine excreted by growing pigs, apparent fecal nutrient digestibility and daily nitrogen balance and retention, meat quality and slaughter value of carcasses was examined. There were no statistically significant differences between the control and experimental group when taking into account the listed parameters. The average value of carcass yield of growing pigs in both groups did not differ from the average carcass yield of pigs from the national population. This indicates that the microalgae meet current standards for carcass yield and can be used by the meat industry. In the algal group, a positive change in terms of reducing fat content in meat, and the growth of the desired protein, was observed. The meat of pigs in the experimental group was characterized by lower water absorption and drip than in the control group. Furthermore, a slight darkening of the meat from pigs in the experimental group was observed. On the basis of conducted experiments, it was found that the enriched algal biomass had no negative effect on the examined parameters and therefore may be introduced in pig nutrition as a feed material as an alternative to inorganic salts.

Author Contributions: Conceptualization, I.M., K.C. and D.K.; methodology, I.M., K.C. and D.K.; formal analysis, I.M., K.C. and D.K.; investigation, I.M., K.C. and D.K.; writing—original draft preparation, I.M.; writing—review and editing, I.M., K.C. and D.K.; visualization, I.M.; supervision, K.C. and D.K.; project administration, K.C.; funding acquisition, K.C. All authors have read and agreed to the published version of the manuscript.

Funding: This research was funded by Polish Ministry of Science and Higher Education, grant number N R05 0014 10.

Conflicts of Interest: The authors declare no conflict of interest.

References

1. Craigie, J.S. Seaweed extract stimuli in plant science and agriculture. *J. Appl. Phycol.* **2011**, *23*, 371–393. [CrossRef]
2. Hurst, D. Marine functional foods and ingredients. A briefing document. *Mar. Fores. Ser.* **2006**, *5*, 1–72.
3. McHugh, D.J. *A Guide to the Seaweed Industry*; FAO Fisheries Technical Paper; Food and Agriculture Organization of the United Nations: Rome, Italy, 2003; p. 441.
4. Makkar, H.P.S.; Tran, G.; Heuzé, V.; Giger-Reverdin, S.; Lessire, M.; Lebas, F.; Ankers, P. Seaweeds for livestock diets: A review. *Anim. Feed Sci. Technol.* **2016**, *212*, 1–17. [CrossRef]
5. Angell, A.R.; Angell, S.F.; de Nys, R.; Paul, N.A. Seaweed as a protein source for mono-gastric livestock. *Trends Food Sci. Technol.* **2016**, *54*, 74–84. [CrossRef]
6. Corino, C.; Modina, S.C.; Di Giancamillo, A.; Chiapparini, S.; Rossi, R. Seaweeds in pig nutrition. *Animals* **2019**, *9*, 1126. [CrossRef]
7. Øverland, M.; Mydland, L.T.; Skrede, A. Marine macroalgae as sources of protein and bioactive compounds in feed for monogastric animals. *J. Sci. Food Agric.* **2019**, *99*, 13–24. [CrossRef]
8. Lynch, M.B.; Sweeney, T.; Callan, J.J.; O'Sullivan, J.T.; O'Doherty, J.V. The effect of dietary *Laminaria*-derived laminarin and fucoidan on nutrient digestibility, nitrogen utilisation, intestinal microflora and volatile fatty acid concentration in pigs. *J. Sci. Food Agric.* **2010**, *90*, 430–437. [CrossRef]
9. Dierick, N.; Ovyn, A.; De Smet, S. Effect of feeding intact brown seaweed *Ascophyllum nodosum* on some digestive parameters and on iodine content in edible tissues in pigs. *J. Sci. Food Agric.* **2009**, *89*, 584–594. [CrossRef]
10. Gardiner, G.E.; Campbell, A.J.; O'Doherty, J.V.; Pierce, E.; Lynch, P.B.; Leonard, F.C.; Stanton, C.; Ross, R.P.; Lawlor, P.G. Effect of *Ascophyllum nodosum* extract on growth performance, digestibility, carcass characteristics and selected intestinal microflora populations of grower–finisher pigs. *Anim. Feed Sci. Technol.* **2008**, *141*, 259–273. [CrossRef]
11. Choi, Y.S.; Choi, J.H.; Han, D.J.; Kim, H.Y.; Kim, H.W.; Lee, M.A.; Chung, H.J.; Kim, C.J. Effects of *Laminaria japonica* on the physico-chemical and sensory characteristics of reduced-fat pork patties. *Meat Sci.* **2012**, *91*, 1–7. [CrossRef]
12. Michalak, I.; Chojnacka, K. Edible macroalga *Ulva prolifera* as microelemental feed supplement for livestock: The fundamental assumptions of the production method. *World J. Microbiol. Biotechnol.* **2009**, *25*, 997–1005. [CrossRef]
13. Michalak, I.; Chojnacka, K.; Witek-Krowiak, A. State of the art for the biosorption process—A review. *Appl. Biochem. Biotechnol.* **2013**, *170*, 1389–1416. [CrossRef]
14. Michalak, I.; Witek-Krowiak, A.; Chojnacka, K.; Bhatnagar, A. Advances in biosorption of microelements—the starting point for the production of new agrochemicals. *Rev. Inorg. Chem.* **2015**, *35*, 115–133. [CrossRef]

15. Jacela, J.Y.; DeRouchey, J.M.; Tokach, M.D.; Goodband, R.D.; Nelssen, J.L.; Renter, D.G.; Dritz, S.S. Feed additives for swine: Fact sheets—high dietary levels of copper and zinc for young pigs, and phytase. *J. Swine Health Prod.* **2010**, *18*, 87–91. [CrossRef]

16. McGrath, S.P.; Chaudri, A.M.; Giller, K.E. Long-term effect of metals in sewage sludge on soils microorganisms and plants. *J. Ind. Microbiol.* **1995**, *14*, 94–104. [CrossRef]

17. Hernández, A.; Pluske, J.R.; D'Souza, D.N.; Mullan, B.P. Levels of copper and zinc in diets for growing and finishing pigs can be reduced without detrimental effects on production and mineral status. *Animal* **2008**, *2*, 1763–1771. [CrossRef]

18. Michalak, I.; Chojnacka, K.; Korniewicz, A. New feed supplement from macroalgae as the dietary source of microelements for pigs. *Open Chem.* **2015**, *13*, 1341–1352. [CrossRef]

19. Dederen, L.H.T. Marine eutrophication in Europe: Similarities and regional differences in appearance. In *Marine Coastal Eutrophication: Proceedings of an International Conference*; Vollenweider, R.A., Marchetti, R., Viviani, R., Eds.; Elsevier: Bologna, Italy, 1992; pp. 21–24.

20. Haroon, A.M.; Szaniawska, A.; Normant, M.; Janas, U. The biochemical composition of *Enteromorpha* spp. from the Gulf of Gdańsk coast on the southern Baltic Sea. *Oceanologia* **2000**, *42*, 19–28.

21. Szefer, P.; Skwarzec, B. Concentration of elements in some seaweeds from coastal region of the southern Baltic and in the Żarnowiec Lake. *Oceanologia* **1988**, *25*, 87–98.

22. Żbikowski, R.; Szefer, P.; Latała, A. Distribution and relationships between selected chemical elements in green alga *Enteromorpha* sp. from the southern Baltic. *Environ. Poll.* **2006**, *143*, 435–448. [CrossRef]

23. Saeid, A.; Chojnacka, K.; Korczyński, M.; Korniewicz, D.; Dobrzański, Z. Effect on supplementation of *Spirulina maxima* enriched with Cu on production performance, metabolical and physiological parameters in fattening pigs. *J. Appl. Phycol.* **2013**, *25*, 1607–1617. [CrossRef] [PubMed]

24. *Feeding Standards for Poultry and Swine*; Polish Academy of Sciences, Institute of Animals Physiology and Feeding: Warsaw, Poland, 2005.

25. Rak, L.; Morzyk, K. *Chemiczne Badania Mięsa*; Wydawnictwo Akademii Rolniczej we Wrocławiu: Wrocław, Poland, 2007. (In Polish)

26. Rutkowski, A. Wartość pokarmowa zbóż dla kurcząt brojlerów. *Rocz. Akad. Rol. Poznaniu* **1996**, *267*, 22–25. (In Polish)

27. Toldrá, F.; Reig, M. Innovations for healthier processed meats. *Trends Food Sci. Technol.* **2011**, *22*, 517–522. [CrossRef]

28. Svoboda, M.; Saláková, A.; Fajt, Z.; Kotrbáček, V.; Ficek, R.; Drábek, J. Efficacy of Se-enriched alga *Chlorella* spp. and Se-enriched yeast on tissue selenium retention and carcass characteristics in finisher pigs. *Acta Vet. Brno* **2009**, *78*, 579–587. [CrossRef]

29. Hintz, H.F.; Heitman, H. Sewage-grown algae as a protein supplement for swine. *Anim. Prod.* **1967**, *9*, 135–140. [CrossRef]

30. Suzuki, K.; Shimizu, Y.; Kadowaki, H. Influence of dried seaweed stalk and breadcrumbs in pig feed on meat production and quality. *Jpn. J. Swine Sci.* **2002**, *39*, 66–70. [CrossRef]

31. Ministry of Agriculture and Rural Development. *Analiza Sytuacji Rynkowej na Podstawowych Rynkach Rolnych w Grudniu 2011 Oraz w Całym 2011 Roku*; Ministry of Agriculture and Rural Development: Warsaw, Poland, 2012. Available online: www.minrol.gov.pl (accessed on 15 May 2012).

32. Sardi, L.; Martelli, G.; Lambertini, L.; Parisini, P.; Mordenti, A. Effects of a dietary supplement of DHA-rich marine algae on Italian heavy pig production parameters. *Livestock Sci.* **2006**, *103*, 95–103. [CrossRef]

33. Olszewski, A. *Technologie Przetwórstwa Mięsa*; WNT: Warsaw, Poland, 2007.

34. Przybylski, W.; Jaworska, D.; Czarniecka-Skubina, E.; Kajak-Siemaszko, K. Ocena możliwości wyodrębniania mięsa kulinarnego o wysokiej jakości z uwzględnieniem mięsności tuczników, pomiaru barwy i pH z zastosowaniem analizy skupień. *Żywność Nauka Technologia Jakość* **2008**, *4*, 43–51. (In Polish)

 Journal of
Marine Science and Engineering

Review

Seaweeds, Intact and Processed, as a Valuable Component of Poultry Feeds

Izabela Michalak [1,*] and **Khalid Mahrose** [2,*]

1   Faculty of Chemistry, Department of Advanced Material Technologies, Wrocław University of Science and Technology, Smoluchowskiego 25, 50-372 Wrocław, Poland
2   Animal and Poultry Production Department, Faculty of Technology and Development, Zagazig University, Zagazig 44511, Egypt
*   Correspondence: izabela.michalak@pwr.edu.pl (I.M.); ostrichkhalid@zu.edu.eg (K.M.);
    Tel.: +48-713202434 (I.M.); +20-1001278452 (K.M)

Received: 1 July 2020; Accepted: 14 August 2020; Published: 18 August 2020

Abstract: Poultry production is an important area of the agricultural economy. Nowadays, there is an interest in novel sources of feed additives that will improve production performance and poultry health. As an easily available and renewable biomass rich in biologically active compounds, seaweeds can meet this demand. Different forms of seaweeds–seaweed powder from naturally occurring biomass, cultivated or waste biomass, extracted compounds, post-extraction residues or liquid extracts–may be used in poultry feeding. Inclusion of this unconventional material in the poultry nutrition can positively influence the poultry performance along with its health and enrich poultry products with active compounds, such as micro- and macroelements, polyunsaturated fatty acids and pigments. Seaweeds also reduce lipids and cholesterol in eggs. Moreover, due to their unique properties, they can serve as an alternative to antibiotic growth promoters. This review presents the latest developments in the use of seaweeds in poultry nutrition, as well as its limitations.

Keywords: poultry; seaweeds; active compounds; poultry performance; health status; food enrichment

---

1. Introduction

Poultry production is an important agricultural subsector in many countries. Poultry are birds which render economic services to humans as a primary supplier of meat, egg and raw materials for different industries (feather, waste products, etc.), source of income and employment to people when compared to other domestic animals [1]. According to USDA (2020), the world chicken meat production in 2020 increased than previous years. In July 2020, the total production of meat reached 100,026 metric tons, whereas in July 2019 it was 99,027 metric tons–an increase of nearly 1%. The demand for poultry meat will increase because in the face of the economic crisis customers are looking for cheaper animal protein. The total world consumption of chicken meat reached 97,908 metric tons in July 2020, whereas in July 2019–97,127 metric tons [2]. Poultry is efficient in converting feed into high-value products within a comparably short period [3–5]. Eggs and poultry meat are beginning to make a substantial contribution to relieving the protein insufficiency in many countries [6,7]. In today's poultry industry, practices regarding management and feeding (composition, systems) are among the most important factors [8–14].

Currently, there is an interest in the application of seaweeds in poultry nutrition. Seaweeds (called also macroalgae), which include green (Chlorophyceae), brown (Phaeophyceae) and red algae (Rhodophyceae), are a naturally occurring source of the biomass that develops in variable environments (results also from eutrophication) and is easily cultivated [15]. Seaweeds as a rich source of bioactive compounds when included into feed can improve poultry health and performance as well as increase the quality of poultry products (eggs, meat) [16,17]. According to the Commission Regulation (EU)

No 575/2011 of 16 June 2011, algae in different forms are listed in the catalog of feed materials, which contains: "algae-live or processed, regardless of their presentation, including fresh, chilled or frozen algae", "dried algae-product obtained by drying algae" that "may have been washed to reduce the iodine content", "algae meal—product of algae oil manufacture, obtained by extraction of algae", "algal oil—product of the oil manufacture from algae obtained by extraction", "algae extract—watery or alcoholic extract of algae that principally contains carbohydrates", "seaweed meal—product obtained by drying and crushing macroalgae, in particular brown seaweed" that "may have been washed to reduce the iodine content". What is important, the name of the feed material should be supplemented by the species.

The literature data show that seaweeds in poultry nutrition are used in both forms: as a feed material and a feed additive. According to the Commission Regulation (EU) No 767/2009 of 13 July 2009 on the placing on the market and use of feed, "feed materials—means products of vegetable or animal origin, whose principal purpose is to meet animals' nutritional needs, in their natural state, fresh or preserved and products derived from the industrial processing thereof and organic or inorganic substances, whether or not containing feed additives, which are intended for use in oral animal-feeding either directly as such or after processing or in the preparation of compound feed or as carrier of premixtures". "Feed additives" according to Regulation (EC) No 1831/2003 of 22 September 2003 on additives for use in animal nutrition are defined as "substances, microorganisms or preparations, other than feed material and premixtures, which are intentionally added to feed or water in order to perform, in particular, one or more of the functions": they "(1) favorably affect the characteristics of feed, (2) favorably affect the characteristics of animal products, (3) favorably affect the color of ornamental fish and birds, (4) satisfy the nutritional needs of animals, (5) favorably affect the environmental consequences of animal production, (6) favorably affect animal production, performance or welfare, particularly by affecting the gastrointestinal flora or digestibility of foodstuffs or (7) have a coccidiostatic or histomonostatic effect". In the European Union feed legislation, intact seaweeds or macroalgae are considered "feed material" not requiring registration, while "extracts" of seaweeds are recognized as "feed additives" requiring an EC authorization act before legal use in animal feeding within the EU.

Macroalgae can be not only a part of the strategy to look for new, natural, ecological and healthy feed materials and/or feed additives, but also for the production of designer poultry products (eggs, meat) enriched with biologically active compounds (e.g., polyunsaturated fatty acids, polyphenols, polysaccharides, pigments, vitamins, amino acids, etc.,) with functional attributes, such as antimicrobial, antioxidant, anti-inflammatory, etc. [16–18]. Consumption of such food can be beneficial to human health. Seaweeds can also be considered as a promising alternative to conventional terrestrial resources used for the production of feed materials/feed additives [19]. Locally available materials, such as seaweeds, can reduce feed cost [20].

In the literature, there are several review articles or book chapters on the use of seaweeds in animal feeding [15,21–25], and a few of them are dedicated to particular species of animals, for example: ruminants (sheep, lambs, goats, cows, calves) [22]; pigs [19,22,23]; rabbits [22]; poultry (broilers, laying hens) [22,26,27]; horses [24]. Literature data confirm that seaweeds can play an important role in the animal feeding, but there is no detailed analysis of the effects of algae in poultry nutrition. This article arrays the current state of knowledge in this field. Appropriately selected seaweeds applied at low inclusion levels can improve not only poultry growth performance and the quality of products, but also their health status (e.g., immune function) due to alteration of gut microbiome and antioxidant properties and can be considered an alternative to antibiotic growth promoters (AGP) used in poultry production [21,28–36]. Most often, seaweeds are used as feed additives for hens and broilers, but there are also a few reports on their application in duck [30,37,38], Japanese quail [39,40] and cockerel [41] feeding.

2. Seaweeds Biologically Active Compounds Important in Poultry Nutrition

The nutritional value of seaweeds is highly variable and depends on many factors such as species, maturity, habitat, geographical origin, area of cultivation, season, harvest time, environmental and physiological variations, water temperature, etc. [19,20,42]. Seaweeds that are most frequently used or recommended for poultry feeding are presented in Figure 1.

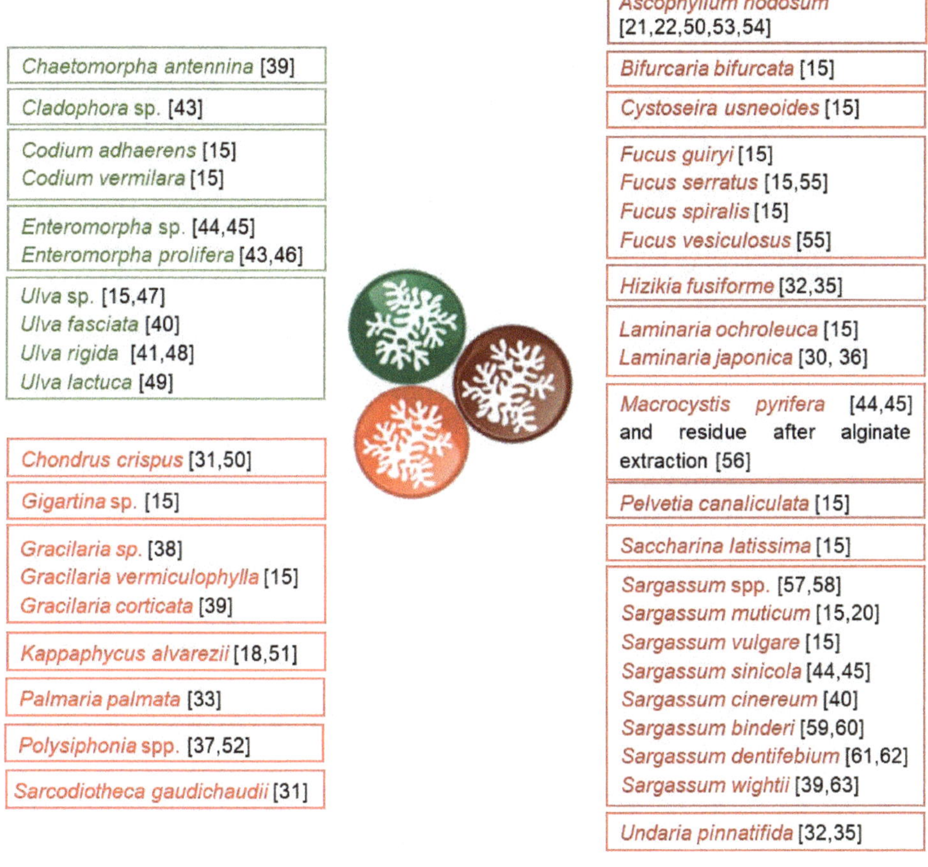

Figure 1. Examples of seaweeds used in poultry feeding [15,18,20–22,30–33,35–41,43–63].

The authors cited in this text presented the chemical composition of individual algae species. As can be seen from Figure 1, brown seaweeds are most frequently applied in poultry feeding. This coincides with the general trend of using mainly brown algae in animal nutrition. Makkar et al. (2016) showed that among main seaweed species used as a component of animal feed, brown—*Ascophyllum nodosum*, *Macrocystis pyrifera*, *Laminaria* and *Sargassum*, red—*Palmaria palmata* and *Lithothamnion* and green species—*Ulva* were dominant [22].

The unique nutritional composition of various seaweed species is of particular importance in poultry nutrition. Seaweeds contain polysaccharides (brown algae: alginate, cellulose, fucoidan, laminarin; red: agar, carrageenan, cellulose, porphyran, xylan; green: cellulose, inulin, pectin, xylan, ulvan), proteins, essential amino acids, minerals (K, Ca, Mg, Zn, Cu, Co, I, B), vitamins (B_{12}, K, C, E, A, D), lipids, polyunsaturated fatty acids (PUFA), pigments such as carotenoids (carotene xanthophyll), chlorophylls, phycobilins (e.g., phycoerythrin) and many antioxidant compounds, including mainly polyphenols [16,17,19]. These bioactive molecules show prebiotic, antimicrobial (antibacterial, antifungal, antivirus), antioxidant, anti-inflammatory immunomodulatory effects [17,23]. Detailed characteristics of many seaweed species are presented in review papers, for example: Fleurence (1999) [64]; Holdt and Kraan (2011) [16]; Cabrita et al. (2016) [15]; Øverland et al. (2019) [19]; Makkar et al. (2016) [22]; Corino et al. (2019) [23].

Seaweeds have great potential to be used as feed additives containing minerals, especially Ca, Mg, Fe, Cu, I, Mn, Se—and to a lesser extent, P and Zn [15,40]. The availability of seaweeds chelated micro-metals to animals (including poultry) is higher than that found in inorganic compounds [21,40]. This observation was also confirmed in the work of Michalak et al. (2011), where microelements bound with green seaweeds via biosorption were better bioavailable to laying hens than inorganic salts of these microelements [43]. Due to the high content of micro- and macroelements, seaweeds are proposed as feed additives that can be involved in the prevention of elemental deficiencies, enrich eggs with minerals, enhance eggshell quality and bone mineralization [43,58]. Pigments contained in seaweeds can be very important in poultry nutrition and are promoted in poultry production due to their antioxidant potential and usefulness for pigmenting food products [65]. Seaweed polysaccharides due to their prebiotic activities can improve poultry performance, egg quality and overall gut health [28,31,34,40,48]. Seaweeds dietary fiber could be an alternative to improve intestinal integrity and reduce concentration of lipids in serum [48]. Marine macroalgae can also serve as a source of polyunsaturated fatty acids in poultry diet and can be an alternative to flaxseed, fish oil, fish meal and canola to enrich eggs and chicken meat with n-3 fatty acids [66]. Carrillo et al. (2008) showed that seaweeds contain several crucial fatty acids such as: linoleic (C18:2, LA), α-linolenic (C18:3, ALA), arachidonic (C20:4, AA), eicosapentaenoic (C20:5, EPA), docosahexaenoic (C22:6, DHA) which are important for human health [44].

On the other hand, seaweeds can also contain toxic metals and other compounds like antinutritional factors (e.g., phenolic compounds, polysaccharides) that can limit their application as feed additives [64,67,68]. Considering the use of seaweeds as feed additives or feed material, the requirements of Commission Regulation (EU) No 1275/2013 of 6 December 2013 concerning the level of undesirable substance (including toxic metals) in products intended for animal feed should be taken into account. Additionally, "feed materials should be free from chemical impurities resulting from their manufacturing process and from processing aids … " according to the Regulation (EC) No 183/2005 of 12 January 2005 laying down requirements for feed hygiene. All seaweed species, before application in animal feeing should be examined in terms of their multielemental composition [42].

Seaweeds polysaccharides can serve as prebiotics, but in some cases they can have an antinutritional effect. It is suggested that certain additives (e.g., enzymes) be incorporated into poultry feed to improve the digestibility of algal polysaccharides [41]. In order to limit the influence of fibers as antinutritional factors, Lahaye and Vigoroux (1992) proposed enzymatic pretreatment of *Palmaria palmata* with commercial endo-β-1,4-d-xylanase to remove polysaccharides [69]; alternatively, bacteria-driven fermentation can be applied to reduce the content of polysaccharides [32,60]. Not all algal species have an antinutritive effect. Ventura et al. (1994) showed that the inclusion of 100, 200 and 300 g of cultivated and then air-dried *Ulva rigida* per kg in the diet of the adult cockerels and 3-week-old chickens did not have an antinutritive effect. Added seaweeds did not modify the true metabolizable energy of the rest of the diet [41]. Another issue which should be considered is the chemical composition of seaweeds which can differ not only by a species, but also due to seasonal variations [19,20,42,64].

3. Forms of Seaweeds in Poultry Feed

In many coastal areas, seaweeds drift to the shore because of sea or ocean waves and become useless waste [59]. Usually, macroalgae are collected manually from surface water, beach and submerged rocks [44]. After drying (e.g., sun, dryer), seaweeds are milled and used as a feed additive [40,44,55,70]. In order to prepare a valuable feed additive, sun drying is not always the best option because it is responsible for the degradation of carotenoids in seaweeds which are known to enhance the color of animal products [44]. Sun-dried seaweeds do not contain vitamin C either because it is extremely sensitive to temperature [71]. Therefore, air drying of seaweeds in shaded areas is advised (e.g., [11,36,72]). The laboratory dryer can also be used for this purpose [52,55], but temperatures should not exceed 40 °C, despite the fact that higher temperatures—for example 60 °C for 72 h—are at times applied [49,52]. Before drying, seaweeds should be washed with fresh water to remove salt, sand particles and epibionts [37,40,48,59]. A high salt content can cause diarrhea and poultry death [59,72]. Dewi et al. (2018) showed that the immersion of *Sargassum binderi* in a flowing river for 15 h enabled a reduction in the salt and ash content and an increase in the organic matter and crude protein content in the algal biomass [59]. As shown in Table 1, dried and ground seaweeds are the most often used form in poultry feeding.

The second source of seaweeds used in feeding experiments on poultry is waste generated from the algae cultivation, which still has a valuable chemical composition. Frasiska et al. (2016) estimated that about 65–70% of the total seaweed cultivation (*Gracilaria* sp. in Indonesia) is waste which has no commercial value and can be converted into feed material for poultry [38]. Ventura et al. (1994) in the feeding experiments on adult cockerels and 3-week-old chicks used green seaweed *Ulva rigida*, which was cultivated, then washed with fresh water, air-dried (~85% dry matter) and finally ground [41].

Before application in poultry feeding, the seaweed biomass is very often subjected to processing, which could positively affect the quality of the feedstuff by improving fiber and nutrient availability [70]. Dewi et al. (2019) proposed fermentation of *Sargassum binderi* with *Bacillus megaterium* S245 (inoculum dosage of 1%) for nine days in order to reduce the content of polysaccharide—alginate. The high content of alginate in brown seaweeds may have a negative effect on the poultry performance: it can bind nutrients and inhibit their absorption in the gastrointestinal tract [60]. Additionally, fermentation of the feedstuff with microorganisms can improve its nutritional quality and elongate the storage period [35,60]. Choi et al. (2014) for fermentation of brown seaweeds—*Undaria pinnatifida* and *Hizikia fusiformis* used five different microorganisms: *Bacillus subtilis*, *Pediococcus acidilactici*, *Pediococcus pentosaceus*, *Saccharomyces cerevisiae* and *Aspergillus oryzae*, among which *B. subtilis* and *A. oryzae* were selected for further research due to the valuable composition of the final feed additive [32]. Because seaweeds are characterized by low digestibility, their fermentation can increase this parameter. Fermentation of seaweeds also enhances their antioxidant, anti-inflammatory and anticoagulant properties [35]. Because of low digestibility, seaweeds and their bioactive compounds (particularly polysaccharides and phenolics) can be treated as prebiotics [48,73], which are defined as "a non-digestible food ingredient that beneficially affects the host by selectively stimulating the growth and/or activity of one or a limited number of bacteria, already established in the colon and thus improves host health" [74]. Seaweed components are resistant to digestion by enzymes present in the gastrointestinal tract and stimulate the growth of beneficial gut bacteria [73].

Besides traditional sun drying, algae can also be boiled or autoclaved as it was presented in the works of Al-Harthi and El-Deek (2012) [70], Al-Harthi and El-Deek (2012) [61] and El-Deek et al. (2011) [62] for *Sargassum dentifebium* and in the work of El-Deek and Al-Harthi (2009) [58] for *Sargassum* sp. It was shown that the method of seaweeds processing had a small effect on their chemical composition. What is important is that thermal processing did not change the *Sargassum* chemical composition and it had absolutely no effect on the content of amino acids and polyunsaturated fatty acids. Even so, it should be taken into account that seaweeds autoclaving, and boiling can change the activity of bioactive compounds. Hence, these seaweed products are not suitable as a replacement of antibiotics. Thermal processing changed the color of *Sargassum*: algae treated that were autoclaved were darker than boiled algae [62]. Each seaweed processing is an added cost, so feed companies would benefit from minimal processing (simple grinding), especially when it has no effect on animal product quality and production parameters [50].

An interesting approach to the preparation of the algal feed additive was proposed by Michalak et al. (2011) [43]. The mixture of marine macroalgae (*Enteromorpha prolifera* and *Cladophora* sp.) was dried and then enriched with microelement ions (Cu(II), Zn(II), Co(II), Mn(II), Cr(III)) via biosorption. The enriched algae were used instead of mineral salts in laying hens feed, which are traditionally used as a source of minerals.

Seaweeds can also be used in poultry nutrition in the liquid form (as extracts) or as extracted compounds and post-extraction residues. Abou El-naga and Megahed (2018) [53] examined the effect of brown seaweed (*Ascophyllum nodosum*) liquid in drinking water (1 mL/L) on broiler chickens' performance and their intestine histology, while Li et al. (2018) tested ulvan extracted from green seaweed (*Ulva* sp.) as a feed additive for laying hens used at doses of 0.05%, 0.1%, 0.5%, 0.8% and 1% [47]. Rendón et al. (2003) proposed to use residue obtained after extraction of alginates from brown seaweed—*Macrocystis pyrifera* (5%) in laying hens feeding. This approach allows the utilization of industrial byproducts [56].

Marine macroalgae used as feed additives can partially substitute for the components of the poultry diet such as corn, sorghum, soybean and mineral salts [44,49]. The maximum level of seaweeds inclusion into the feed strongly depends on their chemical composition, mainly mineral profile including also toxic metals [15]. It is recommended that seaweeds level incorporated into the laying hens' diets should not exceed 10% [44]. Ventura et al. (1994) found that dried *Ulva rigida* was not a suitable ingredient for poultry diets (chicks, cockerels) when the dose was higher than 100 g/kg (10%) [41]. As shown in Table 1, the inclusion level of seaweeds in poultry diet is usually lower than 10%.

4. Enrichment of Poultry Products with Algal Biologically Active Compounds

Functional ingredients of seaweeds can be incorporated into poultry meat and eggs [54]. Many seaweed species are characterized by a high nutrient availability and the rational chemical composition [75]. Enriched with nutrients, animal products can be recognized as functional food [16,43]. Seaweeds can be used to enrich eggs mainly with polyunsaturated fatty acids (e.g., [44,45,66,70]), pigments (enhancement of yolk color) (e.g., [47,55,70]) and minerals (e.g., [43,56]).

In the case of polyunsaturated fatty acids, marine algae can be used as a direct source of dietary n-3 fatty acids (for example docosahexaenoic acid) and can constitute an efficient alternative to currently used sources of these acids (fish oil, fish meal, flaxseed) which are available for the production of poultry products rich in n-3 [76]. Al-Harthi and El-Deek (2012) showed an apparent association between egg yolk fatty acid profile and fatty acid profile of seaweed—palmitic acid was the main saturated fatty acid, whereas oleic—the main unsaturated fatty acid in the biomass of *Sargassum*. The inclusion of 6% of *Sargassum* (used in different forms—sun-dried, boiled or autoclaved) in the diet increased the content of palmitic acid (16:0) in egg yolk, whereas 3% additive increased the content of oleic acid (18:1, n-9) [70]. Carrillo et al. (2008) found that egg yolks from seaweed's groups (*Macrocystis pyrifera*, *Sargassum sinicola* and *Enteromorpha* spp.) presented a higher content of total saturated, monounsaturated and polyunsaturated fatty acids, but the content of the total n-6 fatty acids in all experimental groups was lower and n-3 fatty acids higher than in the control group. Incorporation of especially *M. pyrifera* (10%) in the laying hens' diets is an effective way to increase the content of n-3 fatty acids in egg yolks [44]. Carrillo et al. (2012) showed that these seaweeds can increase the eggs storage time and can protect polyunsaturated fatty acids accumulated in these eggs. Green alga—*Enteromorpha* spp.—had a protective effect on the content of docosahexaenoic acid (DHA) in eggs, while the brown algae—*M. pyrifera* and *S. sinicola*—had a similar effect on the content of eicosapentaenoic acid (EPA). Seaweeds antioxidant compounds such as carotenoids, phenolic compounds and vitamins (C and E) can have an efficient antioxidant effect in eggs enriched with n-3 PUFAs [45]. Mandal et al. (2019) revealed that the feed additive produced from red seaweed—*Kappaphycus alvarezii*—applied in the form of powder at doses 1.25%, 1.50% and 1.75% significantly lowered yolk lipid oxidation in eggs because of its antioxidant properties attributed to active compounds such as polyphenols, β-carotene, vitamins C and E and polysaccharide—carrageenan, which are potential free radical scavengers that inhibit the lipid peroxidation [51]. The polyunsaturated fatty acids (ALA, EPA, DHA) are crucial for normal growth and development and can prevent many diseases such as heart disease, diabetes, hypertension, arthritis, some types of cancer and other autoimmune and inflammatory disorders [65,77]. Since these fatty acids are not synthesized by humans, they should be included in the daily diet [78].

In the poultry industry, seaweeds being a rich source of minerals, natural pigments, polysaccharides (agar, carrageenans, alginic acid, fucoidan, mannitol and laminaran), polyunsaturated fatty acids, sterols (desmosterol, fucosterol, sargasterol, estigmasterol and beta sitosterol) are also used to reduce eggs' cholesterol content according to the requirements of consumers [44,57]. Al-Harthi and El-Deek (2012) reported that seaweeds with their antioxidant properties and the content of fucoxanthin (xanthophyll from brown seaweeds) may reduce the cholesterol level in eggs [70]. These biologically active compounds exhibit hypocholesterolemic and hypolipidemic properties [44,57,79]. This can imply a beneficial impact on human health [70]. Reduction in cholesterol in eggs is important from the point of view of human nutrition: it reduces the risk of cardiovascular disease [47].

There are many reports in literature that confirm the reduction in cholesterol in eggs after seaweeds application. Carrillo et al. (2012) found that decrease in cholesterol level was dose dependent. The cholesterol content in eggs was reduced as the level of *Sargassum* spp. (2%, 4%, 6% and 8%) increased in the hens' diet [57]. Carrillo et al. (2008) showed that among tested seaweeds—*Macrocystis pyrifera*, *Sargassum sinicola* and *Enteromorpha* spp.—supplied at a 10% dose with sardine oil (2%) to the feed of hens, only green macroalga reduced the content of cholesterol when compared to the control group [44]. In the work of Al-Harthi and El-Deek (2012) it was demonstrated that *Sargassum* used in different forms—sun-dried, boiled or autoclaved (3% or 6%) as a feed additive for hens—significantly reduced the cholesterol, high-density lipoprotein (HDL) and triglycerides content in yolks when compared with the control group. A higher decrease was also observed when a higher dose (6%) of different forms of seaweeds was used. The form of seaweeds influenced the content of the total cholesterol and triglycerides in yolk. Generally, the lowest values of these two parameters were in the case of sun-dried seaweeds. For the 6% dose of sun-dried seaweeds, the total cholesterol content in yolk was by 25% lower than in the control group, for boiled seaweeds—by 17% lower and for autoclaved seaweeds by 10% lower. In the case of the triglycerides content in yolk, the decrease for 6% feed additive was as follows: 2.5 times lower for sun-dried than for the control, by 27% lower for autoclaved than for the control and by 61% higher for boiled seaweeds than for the control [70].

Similar results were obtained by Rizk et al. (2017) who observed that the application of dried brown, red and green seaweeds in Sinai hens diet (0.1 and 0.2%) resulted in a decreased content of total lipids (especially red seaweeds), total cholesterol (especially red seaweeds) and low-density lipoprotein (LDL) fraction (especially red seaweeds) and triglycerides (especially red and green seaweeds), while HDL cholesterol was significantly increased (especially red seaweeds) in eggs as compared to control. Seaweeds antioxidant properties are attributed to this effect [80]. Zeweil et al. (2019) demonstrated that sun-dried and ground *Sargassum cinereum* and *Ulva fasciata* used in the feed of laying Japanese quail hens (doses 1.5% and 3%) significantly reduced the content of total lipids and total cholesterol in egg yolk in comparison with the control group, which could be related to their lower levels in blood serum and enhanced total antioxidant capacity [40]. Li et al. (2018) attributed the significant reduction in cholesterol in yolks to polysaccharide ulvan extracted from *Ulva* sp. which was used as a feed additive at concentrations of 0.05%, 0.1%, 0.5%, 0.8% and 1% [47]. The reduction in cholesterol content was probably due to the high content of sulfate in ulvan which has the capacity to decompose cholesterol and has the potential to serve as an antihyperlipidemic agent [79]. Other green macroalga—*Enteromorpha prolifera*—used in the feed of Highland brown laying hens (1%, 2% and 3%) effectively led to a significant reduction in cholesterol content in eggs [46]. In addition, preparations from red seaweeds can reduce free fatty acids and cholesterol in yolks as it was shown in the case of *Kappaphycus alvarezii* used as powder at doses of 1.25%, 1.50% and 1.75% [51]. Zeweil et al. (2019) pointed out that the lessening in egg yolk cholesterol relies on the decline in cholesterol created in the liver. Hence, the decrease in total lipids and total cholesterol can be attributed to the weakening impact of herbal extracts on hepatic 3-hydroxy-3-methylglutaryl coenzyme A (HMG-CoA) reductase that is required to produce cholesterol in the liver [40].

Seaweeds are known to serve as a natural pigment in poultry feed and the enhancement of yolk color is reflected by the deposition of algal carotenoids [65]. Owing to their antioxidant properties, these pigments are beneficial not only for poultry, but also for humans [70]. Lutein, fucoxanthin and zeaxanthin are the main algal carotenoids that improve egg coloring [57]. Al-Harthi and El-Deek (2012) reported that fucoxanthin increased the pigmentation of egg yolks [70]. Contrary results were presented by Strand et al. (1998), who found that fucoxanthin, being the major carotenoid in seaweed meal (*Fucus serratus* and *F. vesiculosus*), was not transferred to yolks of white leghorn laying hens, but gave rise to the following metabolites: fucoxanthinol, fucoxanthinol 3′-sulfate and paracentrone. Generally, the content of carotenoids in egg yolks increased 12–15 times when compared to the control group [55]. Al-Harthi and El-Deek (2012) reported that the inclusion of brown algae (*Sargassum dentifebium*) sun-dried, boiled or autoclaved at doses of 3% and 6% significantly increased the total carotene, lutein plus zeaxanthin content in egg yolks when compared with the control group [70]. Rendón et al. (2003) showed that the residue after alginate extraction from *Macrocystis pyrifera* used at a dose of 5%, can be an interesting alternative to poultry feeding as a natural source of egg pigments (xanthophylls, zeaxanthine, capsolutein) and as a factor that improves the content of egg proteins [56]. Ulvan extracted from *Ulva* sp. applied to the diet of Hy-Line Brown hens at concentrations of 0.5%, 0.8% and 1% for 8 weeks significantly deepen the yolk color into red tendency [47]. The effect of different species of seaweeds on the yolk color is presented in Table 1. Usually, yolk color is determined by means of the Roche yolk color fan, e.g., [35,40,44,55,58] or Chroma Meter through the $L*a*b*$ color system (where $L*$ value is a luminance or lightness component, $a*$ value is a chromatic component from green to red (redness) and $b*$ value is the chromatic component from blue to yellow (yellowness) e.g., [43,65]. For identification of carotenoids in egg yolks of laying hens, Strand et al. (1998) applied several analytical techniques such as HPLC, ^{1}H NMR and UV-vis spectrophotometry [55].

Marine macroalgae can enhance not only the lipid profile of yolks and their color, but also their multielemental composition. Michalak et al. (2011) found that seaweeds—*Enteromorpha prolifera* and *Cladophora* sp.—enriched with microelements—Cu(II), Zn(II), Co(II), Mn(II), Cr(III) via biosorption—increased the content of elements in eggs when compared to the control group, where inorganic salts were used as a source of these elements. This approach was applied to obtain biofortified eggs [43]. Rendón et al. (2003) showed that the addition of 5% of residue from *Macrocystis pyrifera* after alginate extraction to the diet of hens resulted in biofortification of eggs with minerals, especially K, Mg and Fe. Eggs from the experimental group contained also higher crude protein, total lipids, but lower cholesterol levels [56].

Much fewer data are provided about the enrichment of poultry meat with biologically active ingredients derived from seaweeds. Bonos et al. (2016) found that the addition of *Ascophyllum nodosum* (0.5%, 1% and 2%) to the diet of broiler chickens did not influence statistically the total saturated, monounsaturated and polyunsaturated fatty acids in the breast or the thigh meat. Only the group of chickens fed with 2% *A. nodosum* had a significantly higher content of gamma-linolenic fatty acid (C18:3, n-6) in chicken breast meat and a significantly lower of eicosenoic fatty acid content (C20:1, n-9) when compared to the control group [54].

5. Quality of Food Derived from Poultry Feed with Seaweeds

5.1. Egg Quality

Egg quality is a significant criterion for laying hen producers and has essential economic consequences. It results from hens' nutrition and seaweeds used as feed additives can improve egg quality—both physical and biochemical parameters which are crucial for egg producers and consumers. Eggshell quality, which is a visible indicator, is especially important for farmers—higher eggshell weight, thickness and strength will decrease the number of cracked shells. In the case of consumers, eggs with reduced cholesterol content, but also improved yolk color are desired. The egg processing industry requires eggs with intense golden yellow yolks, which are preferred by consumers. Moreover, it is advisable to use natural pigments instead of synthetic carotenoids [58]. Seaweeds can be a solution for such a demand as it was mentioned in the previous section. Beside yolk color, seaweed can also influence other egg quality parameters, the examples of which are presented in Table 1. The percentage increase of a given parameter in the group treated with seaweeds than the control group is presented. It can be clearly stated that seaweeds used as feed additives positively affect yolk color and slightly albumen height and Haugh unit.

In the case of seaweeds, an important issue that should be taken into consideration is egg flavor. Usually, the reduced sensory quality of eggs is reported for hens fed with fish oil. Seaweeds are applied to enhance the content of n–3 fatty acids in yolks and their color to satisfy the consumer's expectation [65]. Carrillo et al. (2008) demonstrated that the inclusion of marine seaweeds—*Macrocystis pyrifera*, *Sargassum sinicola* and *Enteromorpha* spp. (10% plus 2% of sardine oil)—to the feed as a source of n–3 fatty acids did not affect egg flavor [44]. In the work of Herber–McNeill and Van Elswyk (1998), eggs from the experimental groups whose hens were fed with marine macroalgae—2.4% (supplied 200 mg DHA and 69 mg β–carotene/d) and 4.8% (400 mg DHA and 138 mg β–carotene/d)—received acceptable flavor scores, similar to the control group (corn–soybean without added n–3 fatty acids) [65]. Rendón et al. (2003) also proved that egg taste was not affected by the inclusion in poultry diet of the 5% of the post–extraction residue from *Macrocystis pyrifera* [56].

Taking into consideration the physical quality of eggs, the positive effect of different species of seaweeds is especially visible in the case of egg weight and shell thickness. Carrillo et al. (2008) indicated that compounds derived from seaweeds seem to increase egg weight when included in the laying hen diet [44]. Kulshreshtha et al. (2014) fed laying hens with the diet supplemented with red seaweeds, either *Chondrus crispus* or *Sarcodiotheca gaudichaudii* at levels of 0.5%, 1% and 2% and showed that the diet with 1% of *Sarcodiotheca gaudichaudii* boosted egg–yolk and whole egg weights as compared to the control diet; nevertheless, 0.5% and 2% of *Sarcodiotheca gaudichaudii* administration was not diverse from the control. Furthermore, weights of whole eggs and eggshell were greater with hens that ate 1% of *Chondrus crispus* than those which ate 0.5% and 2% of *Chondrus crispus*. The same authors added that egg albumen height, yolk color and shell thickness were not changed due to dietary administration with seaweeds [31]. Choi et al. (2018) found that the supplementation of dry and fermented brown seaweeds (*Undaria pinnatifida* and *Hizikia fusiforme*) to the diet of laying hens (0.5%) improved egg production but did not affect the egg and eggshell characteristics. The significant improvement of eggshell strength of Hy–Line Brown hens can be accounted for by abundant sulfates in ulvan extracted from *Ulva* sp. [47]. Laying Japanese quails fed with diets supplemented with sun–dried and ground green (*Ulva fasciata*) and brown (*Sargassum cinereum*) seaweeds (1.5% and 3%) produced eggs of an improved eggshell thickness, egg yolk weight, index and color when compared to the control and decreased the total lipids and total cholesterol content [40]. Green and brown seaweeds contain different minerals needed for eggshell formation [35]. Wang et al. (2013) reported the improvement of shell thickness of eggs from Highland brown laying hens after the application of *Enteromorpha prolifera* (1%, 2% and 3%), which probably resulted from the increase in calcium in shell [46]. Al–Harthi and El–Deek (2011) found that enzyme supplementation to the diet containing 6% of autoclaved *Sargassum dentifebium* improved eggshell quality while the percentage of shell Ca and P significantly increased [81]. Seaweeds with a high content of easily soluble and digestible calcium can be used at lower concentrations in poultry feeding and can constitute an alternative to limestone, which is the main source of calcium in poultry diet. Higher concentrations of seaweeds can reduce the digestibility of phosphorus, which may lead to reduced skeletal integrity [82]. Bradbury et al. (2012) reported that the high calcium content in calcified seaweeds, included in broilers feed, can improve the bone health status (skeletal integrity) and reduce leg weakness of chickens and their lameness [82].

Table 1. Examples of egg quality parameters of poultry feed with seaweeds. (a) Brown; (b) green: (c) red (in comparison with the control group).

(a) Brown Seaweeds

Form/Inclusion Level	Poultry/Age/ Duration of Experiment	Egg Weight	Yolk Color	Albumen Height	Haugh Unit	Shell Thickness	Shell Weight	Shell Strength	Egg-Shape Index	Ref.
Macrocystis pyrifera, sun-dried and ground; 10% of algae + 2% of sardine oil	Leghorn hens, 35 weeks old, 8-week study	↓ 3.4%	↑ 6.9%	↑ 14%	↑ 8.4%	↓ 7.4%	↓ 9.1%	n.a.	n.a.	[44]
Post-extraction residue from *Macrocystis pyrifera* (after alginate extraction); 5%	Leghorn hens, 23 weeks old, 3-week study	↑ 1.1%	↑ 44%	↑ 5.1%	↑ 0.4%	↑ 3.4%	n.a.	n.a.	n.a.	[56]
Sargassum sinicola, sun-dried and ground; 10% of algae + 2% of sardine oil	Leghorn hens, 35 weeks old, 8-week study	↓ 0.5%	↓ 6.9%	↑ 1.5%	↑ 0.5%	↓ 0.5%	↓ 2.8%	n.a.	n.a.	[44]
Sargassum dentifebium, sun-dried (S); boiled (B) and autoclaved (A) (dried before feeding); 3% and 6%	Hy-Line laying hens, 23 weeks old, 19-week study	n.a.	↑ S: 3%–11%, 6%–4.8%; B: 3%–4.8%, 6%–9.7%; A: 3%–0.9%, 6%–16%	n.a.	n.a.	n.a.	n.a.	n.a.	n.a.	[70]
Sargassum dentifebium, sun-dried (S); boiled (B) and autoclaved (A) (dried before feeding); 3% and 6%	Hy-Line laying hens, 23 weeks old, 19-week study	↑S: 3%–0.2%, 6%–1.0%; ↓B: 3%–1.5%, 6%–3.3%; ↑A: 3%–1.1%, 6%–0.8%	S: 3%↓–2.4%, 6%↑–6.1%; ↑B: 3%–5.5%, 6%–12%; ↑A: 3%–4.3%, 6%–12%	n.a.	↑S: 3%–0.2%, 6%–2.6%; ↑B: 3%–1.8%, 6%–0.7%; ↑A: 3%–0.7%, 6%–1.3%	S: 3%↑–3.3%, 6%↓–2.6%; ↑B: 3%–9.2%, 6%–1.2%; ↑A: 3%–0.5%, 6%–3.3%	n.a.	n.a.	S: 3%↑–0.1%, 6%↓–1.7%; ↓B: 3%–0.9%, 6%–1.7%; A: 3%↓–1.4%, 6%↑–2.0%	[81]
Sargassum cinereum, sun-dried and ground; 1.5% and 3%	Laying Japanese quail hens, 10 weeks old, 14-week study	↑ 1.5%–3.9%; 3%–3.0%	↑ 1.5%–23.1%; 3%–23.1%	n.a.	↓ 1.5%–2.5%; 3%–3.1%	↑ 1.5%–4.6%; 3%–5.1%	↓ 1.5%–2.8%; 3%–2.9%	n.a.	↑ 1.5%–3.2%; 3%–1.6%	[40]
Sargassum spp., dried and ground; 2%, 4%, 6% and 8%	Leghorn hens, 19 weeks old, 5-week study	↓ 2%–1.4%; 4%–2.1%; 6%–0.5%; 8%–2.4%	↑ 2%–3.8%; 4%–4.9%; 6%–9.5%; 8%–12%	↓2%–0.8%; 4%–1.7%; 6%–3.2%; 8%–1.8%	↓ 2%–0.2%; 4%–0.1%; 6%–1.2%; 8%–0.2%	n.a.	n.a.	n.a.	n.a.	[57]

Table 1. *Cont.*

Sargassum spp. sun–dried (S); boiled (B) and autoclaved (A) (dried before feeding); 0%, 3%, 6%, 9% and 12%	Lohman laying hens, 23 weeks old, 19-week study	n.a.	↑*	n.a.	n.a.	n.a.	n.a.	n.a.	n.a.	[58]
Fucus serratus and *F. vesiculosus*, dried at 40 °C and ground; 15%	White Leghorn laying hens, 24 weeks old, 4-week study	n.a.	↑ 12–15 times more carotenoids	n.a.	n.a.	n.a.	n.a.	n.a.	n.a.	[55]
Undaria pinnatifida, 0.5% of seaweed (S) and fermented seaweed (FS)	Hy–Line Brown laying hens, 70 weeks old, 4-week study	S: ↑2%; FS: ↓0.8%	↓ S: 1.4%; FS: 1.4%	n.a.	↓ S: 0.5%; FS: 0.5%	S: ↓25%; FS:-n.c.	n.a.	↓ S: 4.5%; FS: 4.5%	n.a.	[35]
Hizikia fusiforme, 0.5% of seaweed (S) and fermented seaweed (FS)	Hy–Line Brown laying hens, 70 weeks old, 4-week study	S: ↑1.4%; FS: ↓0.3%	↓ S: 1.4%; FS: 1.4%	n.a.	↓ S: 1.8%; FS: 5.4%	S:-n.c.; FS: ↓25%	n.a.	↓ S: 4.5%; FS: 4.5%	n.a.	[35]
Ascophyllum nodosum (Tasco®), sun–dried and ground; 0.25% and 0.5%	Lohmann LSL Brown, 34 weeks old, 3-week study	↑0.25%–1.1%; ↓0.5%– 1.7%	n.a.	↑ 0.25%–2.6%; 0.5%–4.6%	n.a.	↓ 0.25%–0.9%; 0.5%–1.4%	↑0.25%–0.5%; ↓0.5%–2.8%	n.a.	n.a.	[50]
Brown seaweeds (species not defined), dried, 0.1% and 0.2%	Sinai hens, 52 weeks old, 12-week study	↓ 0.1%–4.1%; 0.2%–4.0%	↓0.1%–5.1%; 0.2%–n.c.	n.a.	↓ 0.1%–1.4%; 0.2%–5.4%	↓ 0.1%–16%; 0.2%–7.8%	n.a.	n.a.	0.1%–n.c; ↓0.2%–1.7%	[80]

Table 1. *Cont.*

(b) Green seaweeds

Form/Inclusion Level	Poultry/Age/ Duration of Experiment	Egg Weight	Yolk Color	Albumen Height	Haugh Unit	Shell Thickness	Shell Weight	Shell Strength	Egg–Shape Index	Ref.
Enteromorpha spp., sun–dried and ground, 10% of algae + 2% of sardine oil	Leghorn hens, 35 weeks old, 8-week study	↓ 1.4%	↓ 19%	↓ 12%	↓ 9.3%	↓ 1.6%	↓ 3.6%	n.a.	n.a.	[44]
Enteromorpha prolifera, 1%, 2% and 3%	Highland brown laying hens, 42 weeks old, 4-week study	↑ 1%–8.7%; 2%–6.0%; 3%–8.1%	↑ 1%–19%; 2%–18%; 3%–25%	n.a.	n.a.	↑ 1%–2.9%; 2%–2.9%; 3%–5.7%	n.a.	n.a.	↑ 1%–0.8%; 2%–n.c.; 3%–n.c.	[46]
Ulvan from *Ulva* sp., 0.05%, 0.1%, 0.5%, 0.8% and 1%	Hy–Line Brown hens, 61 weeks old, 8-week study	↓0.05%–0.9%; ↑0.1%–0.7%; ↓0.5%–0.1%; ↓0.8%–0.2%; ↓1%–1.2%	↑ *	↓ *	↓ *	n.a.	n.a.	↑ *	↓ *	[47]
Ulva fasciata, sun–dried and ground; 1.5% and 3%	Laying Japanese quail hens 10 weeks old, 14-week study	↑ 1.5%–3.9%; 3%–2.6%	↑ 1.5%–35%; 3%–46%	n.a.	↓ 1.5%–2.3%; 3%–2.3%	↑ 1.5%–3.5%; 3%–4.1%	↓ 1.5%–2.8%; 3%–1.2%	n.a.	↑ 1.5%–2.5%; 3%–0.9%	[40]
Green seaweeds (species not defined), dried, 0.1% and 0.2%	Sinai hens, 52 weeks old, 12-week study	↑ 0.1%–4.4%; 0.2%–0.9%	↓0.1%–5.1%; 0.2%–n.c.	n.a.	↓ 0.1%–2.7%; 0.2%–5.4%	↓ 0.1%–9.9%; 0.2%–16%	n.a.	n.a.	↑ 0.1%–1.8%; 0.2%–0.9%	[80]
Enteromorpha prolifera and *Cladophora* sp. enriched with microelements (S)– Cu, Mn, Zn, Co, Cr	Lohmann brown laying hens, 30–45 weeks old, 5-week study	↑ S–Cu–3.1%; S–Mn–5.7%; S–Zn–2.0%; S–Co–8.7%; S–Cr–1.6%	↑ *	n.a.	n.a.	↑ S–Cu–10%; S–Mn–14%; S–Zn–6.9%; S–Co–7.4%; S–Cr–5.4%	n.a.	n.a.	n.a.	[43]

Table 1. *Cont.*

(c) Red Seaweeds

Form/Inclusion Level	Poultry/Age/ Duration of Experiment	Egg Weight	Yolk Color	Albumen Height	Haugh Unit	Shell Thickness	Shell Weight	Shell Strength	Egg–Shape Index	Ref.
Red seaweeds (species not defined); dried, 0.1% and 0.2%	Sinai hens, 52 weeks old, 12-week study	↑0.1%–1.2%; ↓0.2%–7.7%	↓ 0.1%–5.1%; 0.2%–5.1%	n.a.	↓0.1%–6.4%; 0.2%–n.c.	↓ 0.1%–9.0%; 0.2%–7.8%	n.a.	n.a.	↑ 0.1%–0.5%; 0.2%–0.5%	[80]
Chondrus crispus, raw ground (G: 4%) and extruded, dried and reground (E: 1%, 2%, 3% and 4%)	Lohmann Brown Lite laying hens, 70 weeks old, 3-week study	G: ↑4%–0.4%; E: ↓1%–1.6%; E: ↓2%–2.0%; E: ↑3%–1.1%; E: ↑4%–0.3%	n.a.	G: ↑4%–4.2%; E: ↑1%–6.9%; E: ↑2%–2.8%; E: ↑3%–8.3%; E: ↓4%–2.8%	n.a.	G: ↓4%–2.0%; E: 1%–; E: ↑2%–2.0%; E: 3%–; E: ↑4%–4.0%	G: ↓4%–0.2%; E: ↓1%–0.5%; E: ↓2%–0.8%; E: ↑3%–0.5%; E: ↑4%–2.9%	n.a.	n.a.	[50]
Kappaphycus alvarezii, powder; 1.25%, 1.50% and 1.75%	Laying hens, day–old, 40-week study	↑ 1.25%–0.7%; 1.50%–3.8%; 1.75%–5.3%	↓ 1.25%–10%; 1.50%–6.6%; 1.75%–12%	n.a.	↑ 1.25%–0.6%; 1.50%–4.9%; 1.75%–5.3%	↑ 1.25%–12%; 1.50%–9.1%; 1.75%–12%	n.a.	n.a.	↑ 1.25%–0.3%; 1.50%–4.7%; 1.75%–5.2%	[51]
Marine macroalgae (not defined); 2.4% and 4.8%	Single Comb White Leghorn, 56 weeks old, 4-week study	n.a.	↑ *	n.a.	n.a.	n.a.	n.a.	n.a.	n.a.	[65]

n.a.—not available, n.c.—no change, ↑—increase, ↓—decrease in comparison with the control group, * numeric values not available.

5.2. Carcass Characteristics and Meat Quality

In poultry production, broiler chickens are selected for rapid growth, heavier breast weight and increased muscle mass. Many papers examined the effect of seaweeds incorporation into the diet of broilers on their production performance and carcass traits (e.g., [18,20,28,29,32,33,36,48,49,53,62,63,75,83–88]). Generally, seaweeds positively influence meat quality, which is usually improved as a consequence of the reduction in fats.

Zahid et al. (2001) showed that broiler chickens (Hubbard) fed on the normal feed containing brown seaweeds (10%, 20%, 30% and 40%) had higher body weight and a lower amount of fat while higher of protein as compared to controls [83]. Similar results were reported also for brown seaweed—*Ascophyllum nodosum*—which was added as a liquid to drinking water (1 mL/L) of Indian River chicks and significantly increased dressing color, breast width and length and decreased body fat when compared to the control chicks [53]. Much work has been dedicated to other brown seaweed—*Sargassum* sp. Erum et al. (2017) found that with the increasing level of air–dried and ground *Sargassum muticum* in the feed, fat pads of the birds were reduced. The carcass quality was improved due to the reduction in fat. In the control group (without algae) and group with 5% inclusion of *S. muticum*, the color of fat pads and meat was yellowish. With the increase in *S. muticum* level till 10%, fat pads were yellowish but minimal, whereas meat—slightly reddish. For the highest dose of *S. muticum*—15%—there were no fat pads and the meat was reddish [20]. In the work of Athis Kumar (2018), the consumers evaluated meat parameters such as color, flavor, tenderness, juiciness and taste. Meat from broilers fed with basal diet with the addition of 1% or 2% of *Sargassum wightii* received the highest score. The inclusion of this brown seaweed (1%, 2%, 3% and 4%) into broiler diet caused the enhancement of carcass traits such as: weight of legs, breast, thigh and dressing. The maximum supplementary effect was noted for *Sargassum* doses of 1% and 2% and was attributed to the chemical composition of this seaweed that contains minerals, vitamins, polyunsaturated fatty acids, essential amino acids, sterols and polysaccharides such as fucoidan [63]. El–Deek et al. (2011) found that different levels (2%, 4% and 6%) of thermally processed *Sargassum dentifebium* (boiled, autoclaved) in broiler finisher diets had an insignificant effect on the dressing percentage than raw (untreated) algae [62].

In the case of green seaweeds, Wang et al. (2013) showed that the addition of 2%, 3% or 4% of dry algal powder from *Enteromorpha prolifera* improved the breast meat quality: the content of fat was significantly reduced, the thickness of subcutaneous fat and abdominal fat rate were decreased [87]. Another green seaweed—*Ulva lactuca*—was tested as a feed additive in the work of Abudabos et al. (2013). Broiler chickens (male chicks—Ross) which received 3% of the algal additive had the highest breast muscle yield and dressing percentage whereas abdominal fat was significantly reduced when compared with the group that received the 1% dose and the control group. The color of breast muscle was not affected by any dietary treatments. Improvement in dressing and breast yield can attributed to the higher content of crude protein and amino acids, specially methionine in the feed with seaweeds [49]. Contrary results were presented by Cañedo–Castro et al. (2019), who showed that there were no significant differences in carcass weight and yield of Arbor Acres broilers fed with different levels of *Ulva rigida* (air dried and ground) inclusion in the diet—2%, 4% and 6% [48].

In the case of red seaweed—*Kappaphycus alvarezii*—added to the diet of broiler chickens at levels of 0.25%, 0.5%, 0.75%, 1%, 1.25%, and 1.5% there occurred the enhancement of live weight, carcass traits (dressing and eviscerated percentage) and organ weight (heart, liver, gizzard) [18]. A similar effect was also reported for other red seaweeds (*Sarcodiotheca gaudichaudii*, *Chondrus crispus*) used as powder in layer hens at doses of 0.5%, 1% or 2%: no significant effects on liver, spleen, ileum and heart weight [31]. Red algae powder (*Chondrus crispus*) applied in a feed of male broiler chickens (0.3%) significantly improved the carcass and breast yield and decreased the abdominal fat yield [88]. Red seaweed—*Polysiphonia* spp.—was also introduced to the diet of ducks at doses of 5%, 10% and 15% and the highest level did not significantly affect the relative weight of dressing, thigh and breast muscles. The relative weight of breast muscles significantly increased when seaweeds were applied at 5% and 10%. Seaweed at a dose of 15%, significantly improved the texture of breast muscles and at doses of 5% and 10% improved the texture of thigh muscles. There were no significant differences in the taste, aroma, juiciness and color of meat [37].

6. Effect of Seaweeds on Poultry Growth and Productive Performance

6.1. Growth Performance

Growth performance in poultry includes live body weight at marketing (BW), average of daily gain (ADG), feed consumption (FC), feed conversion ratio (FCR) and mortality rate. Seaweeds influence growth performance parameters, the examples of which are presented in Table 2. Fermented by–products of brown seaweeds (*Undaria pinnatifida*) and seaweed fusiforme (*Hizikia fusiformis*) used at a dose of 0.5% caused high ADG and FCR and a low mortality rate of broilers when compared with the control group [32]. Erum et al. (2017) indicated that feeding broiler chickens on marine macroalga (*Sargassum muticum*) as a feed additive (5%, 10% and 15%) led to the improvement in BW at marketing, ADG, FC and FCR [20]. The authors indicated that ADG of broilers increased in proportion to the increase in the substitution level of algae; however, birds that received 10% of algae had the lowest FCR. Athis Kumar (2018) fed broiler chicks on a diet supplemented with *Sargassum wightii* powder (1%, 2%, 3% and 4%) and observed that FCR was boosted from 51.5% to 51.8% rather than 33.9% in the control, while the weight gain was higher in broilers fed with the diet supplemented with 4% of *Sargassum wightii* powder, by about 51% in comparison with the control group. The author attributed the beneficial effects of *Sargassum wightii* powder to its palatability, high content of nutrients and its ability to enhance digestion and absorption of nutrients in the gut [63]. Bai et al. (2019) indicated that, when compared with the control, dietary supplementation of *Laminaria japonica* powder (1%) improved FCR in broiler chicks due to the increased dietary energy content [36]. *Ulva rigida* was used as a prebiotic in broiler diets (2%, 4% and 6%) to enhance growth performance [48]. Non–significant variations in BW were observed, but FC, FCR and the mortality rate presented significant alterations. Feed consumption was greater in broilers that consumed 4% and 6% of the prebiotic owing to its attractant properties, while mortality was higher in the control group and in the group of broilers fed with the addition of 6% of *Ulva rigida*. Feeding of laying hens (30–45 weeks of age) on diets supplemented with two marine macroalgae (*Enteromorpha prolifera* and *Cladophora* sp.) enriched with microelements (Cu, Zn, Co, Mn and Cr) resulted in a rise in BW of hens [43]. On the other hand, Islam et al. (2014) stated that ADG and FCR of growing ducks fed on a diet supplemented with sea tangle (*Laminaria japonica*) did not significantly differ from the control group [30]. In the same line, El–Deek and Brikaa (2009) reported that growing ducks fed with diets supplemented with 4%, 8% and 12% of red seaweeds (*Polysiphonia* spp.), irrespective of the diet form (pellet or mash), showed non–significant differences in BW, ADG, FC and FCR [37]. When Japanese quails were fed with the diet supplemented with dried seaweeds (3%) harvested from the Gulf of Mannar, India (*Chetomorpha antennina*, *Sargassum wightii* and *Gracilaria corticata*), BW and FCR still remained without change [39]. Substituting 1% and 3% of corn with green seaweed (*Ulva lactuca*) in broiler diets resulted in insignificant differences in BWG, cumulative FC and FCR [49].

Table 2. Examples of growth performance of poultry fed diets with seaweeds (in comparison with the control group).

Seaweeds	Poultry Species	Growth Performance Parameters				
		Final Body Weight	Body Weight Gain	Feed Intake	Mortality Rate	Ref.
Brown Seaweeds						
Undaria pinnatifida, 0.5% of seaweed (S) and fermented seaweed (FS)	Broilers, one day old Ross male, 35-day study	↑ (35 day) S–4.3%; FS–2.5%	↑ (35 day) S–7.6%; FS–6.4%	(35 day) ↑S–2.4%; ↓FS–2.7%	↓ S– ~4.5 times FS– ~9 times	[32]
Hizikia fusiformis, 0.5% of seaweed (S) and fermented seaweed (FS)	Broilers, one day old Ross male, 35-day study	↑ (35 day) S–2.2%; FS–0.7%	↑ (35 day) S–4.1%; FS–3.2%	↓ (35 day) S–1.3%; FS–1.5%	↓ S–3 times FS– ~9 times	[32]
Sargassum muticum, air dried under the shade and ground; 5%, 10%, 15%	Broilers, one day old, 39-day study	↑ (39 day) 5%–23%; 10%– 25%; 15%–26%	↑ (39 day) 5%–25%; 10%–27%; 15%–28%	↑ (39 day) 5%–20%; 10%–14%; 15%–17%	n.a.	[20]
Sargassum wightii, dried under shade, then sun–dried and ground, powder; 1%, 2%, 3%, 4%	Broilers, one day old, 121-day study	↑ 1%–88%; 2%–93%; 3%–93%; 4%–93%	↑ 1%–99%; 2%–104%; 3%–104%; 4%–104%	↑ 1%–53%; 2%–58%; 3%–58%; 4%–58%	n.a.	[63]
Sargassum wightii, dried; 3%	Japanese quail, one day old, 42-day study	↓ (42 day) 3%–0.3%	n.a.	n.a.	n.a.	[39]
Sargassum dentifebium, sun–dried; 2%, 4%, 6%	Broilers, 18 days old, 39-day study	↓ (39 day) 2%–1.3%; 4%–2.7%; 6%–3.3%	↓ (18–39 day) 2%–1.5%; 4%–3.7%; 6%–5.0%	↑ (18–39 day) 2%–1.2%; 4%–2.5%; 6%–5.9%	n.a.	[62]
Laminaria japonica, commercial powder and charcoal; 0.1%, 0.5%, 1%	Duck, 22 days old, 21-day study	n.a.	(0–21 days) 0.1%–n.c.; 0.5%–n.c.; ↑1%–4.8%	↓ (0–21 days) 0.1%–0.8%; 0.5%–2.7%; 1%–1.6%	n.a.	[30]

Table 2. *Cont.*

Seaweeds	Poultry Species	Growth Performance Parameters				
		Final Body Weight	Body Weight Gain	Feed Intake	Mortality Rate	Ref.
Laminaria japonica, commercial powder; 1%	Arbor Acres broilers, one day old, 42-day study	n.a.	↑ 1%–2.7%	↓ 1%–0.02%	n.a.	[36]
Red Seaweeds						
Polysiphonia spp., dried; 1.5%, 3%	Hubbard duck, 14 days old, 56-day study	↓ (56 day) 1.5%–1.3%; 3%–3.6%	↓ 1.5%–1.3%; 3%–3.8%	↓ 1.5%–1.7%; 3%–3.0%	n.a.	[52]
Gracilaria corticata, dried; 3%	Japanese quail, one day old, 42-day study	↑ (42 day) 3%–0.05%	n.a.	n.a.	n.a.	[39]
Kappaphycus alvarezii, 0.25%, 0.5%, 0.75%, 1%, 1.25%, 1.5%	Broilers, one day old, 42-day study	n.a.	↑ (0–42 days) 0.25%–1.8%; 0.5%–2.4%; 0.75%–3.0%; 1%–2.9%; 1.25%–11%; 1.5%–9%	(0–42 days) ↑0.25%–1.3%; ↓0.5%–2.6%; ↓0.75%–1.4%; ↓1%–0.3%; ↑1.25%–4.8%; ↑1.5%–2.5%	n.a.	[18]
Palmaria palmata, dried, ground, commercial; 0.6%, 1.2%, 1.8%, 2.4%, 3%	Ross 308 broilers, one day old, 35-day study	(25–35 day) ↑0.6%–2.0%; ↑1.2%–1.0%; ↑1.8%–5.3%; ↓2.4%–1.7%; ↓3%–2.7%	(25–35 day) ↓0.6%–12%; ↓1.2%–9.5%; ↓1.8%–9.8%; ↓2.4%–3.6%; ↓3%–14%	(25–35 day) ↓0.6%–4.5%; ↓1.2%–0.1%; ↓1.8%–7.0%; ↓2.4%–8.1%; ↓3%–5.4%	(0–35 day) 0.6%–lack mortality; ↓1.2%–33%; 1.8%–lack mortality; ↓2.4%–67%; ↓3%–67%	[33]

Table 2. *Cont.*

Seaweeds	Poultry Species	Growth Performance Parameters				
		Final Body Weight	Body Weight Gain	Feed Intake	Mortality Rate	Ref.
Green Seaweeds						
Ulva rigida, dried under shade, ground; 2%, 4%, 6%	Arbor Acres Broilers, one day old, 42-day study	(42 day) ↑2%–0.8%; ↓4%–5.5%; ↑6%–0.6%	n.a.	↑ (42 day) 2%–1.9%; 4%–2.6%; 6%–4.6%	↓ 2%–4 times; 4%–4 times; 6%–17%	[48]
Chetomorpha antennina, dried; 3%	Japanese quail, one day old, 42-day study	↑ (42 day) 3%–0.1%	n.a.	n.a.	n.a.	[39]
Ulva lactuca, sun–dried and then oven–dried, ground; 3%, 6%	Ross broilers, one day old, 33-day study	n.a.	(19–33 days) ↓3%–0.4%; ↑6%–2.3%	(19–33 days) ↓3%–0.9%; ↑6%–2.7%	n.a.	[49]

n.a.—not available, n.c.—no change, ↑—enhancement, ↓—decrease.

6.2. Egg Production Performance and Hatchability

Dietary administration of red seaweed either *Chondrus crispus* or *Sarcodiotheca gaudichaudii* at levels of 0.5%, 1% and 2% had a significant impact on hen–day egg production as found by Kulshreshtha et al. (2014). Hen–day egg production was greater in hens fed with the diet with 2% of *Sarcodiotheca gaudichaudii* and 1% of *Chondrus crispus* than the control hens [31]. Egg production rate and egg mass were superior ($p < 0.05$) in laying hens that consumed diet supplemented with brown seaweed (*Undaria pinnatifida*) than those fed with the control or fermented (*Undaria pinnatifida*) or with non–fermented seaweed (fusiforme—*Hizikia fusiformis*)–supplemented diets. The hens fed with seaweed fusiforme–supplemented diet and those that ate fermented seaweed fusiforme–supplemented diet had a greater egg production ($p < 0.05$) than the control group (Choi et al. 2018), as a consequence of the high content of brown algae polysaccharides which improved laying performance and immune status [35]. The latter authors also indicated that dietary brown algae inclusion was able to enhance egg–laying performance and the supplementation with fermented seaweeds had no helpful impact on the egg–laying performance.

Egg laying rate in hens was enriched by 4.4% and 4.3% as compared to the control when hens consumed diets supplemented with red and green seaweeds (0.1 g/kg diet), during 52–64 weeks of age, thanks to the valuable constituents observed in seaweeds [80]. Ulvan extracted from green seaweed—*Ulva*—was supplemented in the diet of Hy–Line Brown laying hens (61 weeks old). The findings of this work showed that ulvan at levels from 0.1 to 1% can enhance egg production, egg weight and eggshell strength [47]. Bratova and Ganovski (1982) stated that consuming diets supplemented with black sea algae had an encouraging impact on hatchability. The maximum percent of hatchability—87.95%—was a 6.85% increase on the control group and can be attributed to the addition of 2% seaweed to poultry feed [89]. Zeweil et al. (2019) concluded that supplementing dried green and brown seaweeds brought about significantly greater fertility and hatchability percentages than those of the control group [40].

In the study performed by Zeweil et al. (2019), the laying rate of Japanese quails was enhanced by 8.8%, 7.2% and 11.4, 9.0% for birds fed with the diet administrated with green or brown seaweeds at levels of 1.5% and 3%, correspondingly. The authors also indicated that the same levels of dietary green and brown seaweeds enhanced the hatching rate by 13.3%, 16.1%, 7.2% and 15.2% of total eggs and improved the weight of newly–hatched chicks by 6.4%, 14.1%, 9.9% and 12.8%, correspondingly [40]. Dietary incorporation of two marine macroalgae (*Enteromorpha prolifera* and *Cladophora* sp.) enriched with microelements in laying hens diet increased the number of eggs in the experimental groups when compared to the control [43].

7. Effect of Seaweeds on Poultry Health

Among the huge number of materials that are the source of bioactive compounds in the diet of poultry, seaweeds are a valuable and readily available resource [90] so much so that they are known to influence positively the poultry health. Seaweed extracts have antimicrobial and antiviral properties along with immunomodulatory influences [31,32,34,36]. Seaweeds could also be used as prebiotics for improving the production and health status of poultry species [48]. The supplementation of *Ulva rigida* to broilers diet enhanced the growth of intestinal villi and decreased serum total cholesterol and triglyceride concentrations and can be treated as a prebiotic that can improve broiler health [48]. Polysaccharides are the greatest well–recognized complexes in seaweeds, which underwent multifaceted investigations as a result of their broad bioactivities. The antiviral properties of seaweeds derive from the presence of such bioactive compounds as carrageenan, alginate, fucan and laminaran [91]. Furthermore, these bioactive compounds can prevent the joining or internalization of the virus into the host cells or control DNA repetition and protein production [91,92]. Elizondo–Gonzalez et al. (2012) demonstrated that fucoidan (a sulfated polysaccharide existing in the cell wall matrix of *Cladosiphon okamuranus*) displayed action against Newcastle disease virus. Fucoidan works in the initial periods of viral infection so as to prevent viral–induced syncytia creation, possibly by blocking the F protein, which is responsible for fusion of cell membrane and the viral envelope and through conformational modifications [92]. Villus measurements (width, height and contour length) were higher when *Ulva rigida* meal was incorporated into broiler diet (2%, 4% and 6%) than in the control [48]. Intestinal villus width ranged between 0.6 and 0.7 mm in broilers which consumed *Ulva rigida* when compared to the control group (0.4 mm). The highest values of intestinal villus height (1.6 mm) and villus contour length (3.4 mm) were observed in the group fed with 2% *Ulva rigida*.

Dietary supplementation of red seaweed—*Chondrus crispus*—constrained the settlement of *Salmonella* in the excreta and ceca and this could be caused by an increase in the development of *Lactobacillus* and raising the level of short chain fatty acids [34]. The same authors also indicated that a greater concentration of IgA in birds fed on diets complemented with *Chondrus crispus* confirmed the role of macroalgae in the maturation of the immune system. Bai et al. (2019) found that *Laminaria japonica* powder and antibacterial peptide (cecropin) could be applied as diet inclusion for boosting the immune system of broilers. The dietary inclusion of 3% of *Laminaria japonica* powder together with 300 mg/kg of cecropin enhanced the number of serum Newcastle disease antibody titers and lymphocyte during the fattening period of broilers. The bioactive compounds of *Laminaria japonica* stimulated lymphocytes, altered their cell structure, which influenced immunity [36]. Choi et al. (2014) concluded that the diet supplemented with 0.5% of fermented by–products of brown seaweed (*Undaria pinnatifida*) and seaweed fusiforme (*Hizikia fusiformis*) activated broiler humoral immunity and supported physical health. Significantly higher concentrations of IgA and IgM and lower of IgG in the serum of broilers fed with seaweeds than in the control group suggested the effectiveness of the feed additive used [32].

Curiously, the nutritional constituents of *Laminaria japonica* powder may enhance the duplication of *Lactobacillus* and improve the intestinal microecological setting [93,94]. The inclusion of red seaweeds (*Chondrus crispus, Sarcodiotheca gaudichaudii*; 2% or 4%) to the diet of laying hens increased the comparative number of helpful microorganisms (*Bifidobacterium longum, Lactobacillus acidophilus* and *Streptococcus salivarius*) and reduced the pathogenic microbes (*Clostridium perfringens*) in the ileal fillings [31,34]. The administration of *Laminaria japonica* powder mixed with cecropin (extracted from silkworm) extremely constrained *Escherichia coli* intensification and boosted *Lactobacillus* development [36]. *Laminaria* spp. can also enhance humoral immune protection against pathogens [36,95].

Seaweeds can also be used as a feed additive rich in calcium, which can be useful in the treatment of poultry leg weakness and lameness. Supplementation of the broiler diet with a highly digestible marine calcium source, at lower dietary concentration (0.6%) may prevent reduced skeletal integrity [82]. Poultry nutrition and the litter quality are the main factors that are responsible for the development of foot pad dermatitis (PFD), which is characterized by ulcerated lesions on the pad of the foot [9,96]. Abd El–Wahab et al. (2018) examined the effect of different protein sources (soybean, rapeseed, hemoglobin and algae meal) on the FPD score in broilers (160 one day old). Foot pad dermatitis severity was significantly higher ($p < 0.05$) in birds fed with rapeseed and algae meal in comparison with those fed with soybean or hemoglobin meal. The high FPD score can result from the chemical composition of diet (the high content of sodium—about 2 g/kg of dry mass and potassium—about 9 g/kg of dry mass) which caused excessive consumption of water. It affected the deterioration of the litter quality [96].

8. Effect of Seaweeds on Blood Profile

In broilers that ate diet enriched with *Sargassum wightii* powder (1%, 2%, 3% and 4%), the serum concentration of glucose was augmented from 206.1 mg/dL (for the dose of 1%) to 208.7 mg/dL (4%) when compared to 204.2 mg/dL in the control group; total proteins marginally rose from 2.2 mg/dL in the control group to about 2.3 mg/dL in all experimental groups. Albumin level increased from 1.7 mg/dL (1%) to 1.92 mg/dL (3%) when compared to 1.1 mg/dL in the control group and the concentration of globulin decreased from 1.1 mg/dL in the control group to 0.41 mg/dL for the dose of 4%. Triglycerides increased from 96.4 mg/dL (1%) to 113.2 mg/dL (4%) instead of 80.3 mg/dL in the control; while the cholesterol level was reduced from 122.1 mg/dL in the control group to 96.7 mg/dL (3% and 4%). In the case of macroelements, phosphate concentration increased from 7.6 mg/dL (1%) to 8.6 mg/dL (4%) when compared to 5.5 mg/dL in the control group and the calcium level was improved from 11.3 mg/dL (1%) to 12.2 mg/dL (3% and 4%) instead of 10.4 mg/dL in the control [63]. Using two dietary levels (3% and 6%) of brown marine alga (*Sargassum dentifebium*) processed by various techniques (sun–drying, boiling, autoclaving) reduced the plasma total cholesterol level and lowered the density of the low density lipoprotein (LDL) in laying hens (23–40 weeks of age) than the control group. This decrease was due to the chemical composition of *Sargassum dentifebium*, which is rich in fiber, sterols and other bioactive compounds with antioxidant properties [70]. The dietary supplementation of 3% of *Laminaria japonica* powder combined with 300 mg/kg cecropin increased lymphocyte numbers in broilers [36].

Frasiska et al. (2016) investigated the impact of the diet containing *Gracilaria* sp. waste (10%, 12.5% and 15%) with multi–enzyme additives on lipid profiles of duck (22 weeks old) blood [38]. The authors found that diet with 12.5% of *Gracilaria* sp. plus multi–enzyme additive significantly decreased triglycerides, LDL and cholesterol level, but increased the concentration of high density lipoprotein (HDL) in blood, what can be attributed to *Gracilaria* sp. fiber, which operates as an anticoagulant, antihyperlipidemic, antitumor, antiviral and anti–cholesterol agent [97]. Using 2% of *Ulva rigida* meal as a dietary supplementation in broiler diets caused 10%–14% and 15%–28% reductions in the total cholesterol and triglyceride concentrations, respectively, than the control, due to chemical complexes present in algae like, polysaccharides, sterols and polyunsaturated fatty acids [48]. In an experiment conducted by Rizk et al. (2017), supplementation of hen diets with green, brown and red seaweed (0.1% and 0.2%) lowered the concentration of serum total cholesterol, LDL, very low density lipoprotein (VLDL), triglycerides and total lipids, amplified HDL cholesterol and improved serum total protein, albumin, globulin, liver and kidney functions than the control group. The same authors added that hens fed with the diet supplemented with 0.2% of brown seaweed had significantly higher serum total antioxidant capacity (TAC) and the concentration of glutathione peroxidase (GPX) enzyme, but supplementing the hen diet with 0.2% of red seaweed significantly increased the activity of serum superoxide dismutase (SOD) enzyme [80]. Similar results were also obtained when Zeweil et al. (2019) fed laying Japanese quail with diets containing green and brown seaweeds at levels of 1.5% and 3% [40]. In laying hens fed with the diet enriched with fermented and non–fermented brown algae, Choi et al. (2018) observed that the supplemented–diet groups contained higher albumin concentrations than the control group and the fermented brown seaweed group encompassed higher total cholesterol and triglyceride levels than the other groups. The authors also found that brown seaweed and fermented brown seaweed groups were characterized by upper glutamic pyruvic transaminase activities than the other groups [35].

Abudabos et al. (2013) indicated that serum enzymes and electrolytes were not impacted by the dietary inclusion (1% and 3%) of green algae (*Ulva lactuca*) except for alanine transaminase activity, which was lesser for the groups under consideration [49]. Microelements (Cu, Mn, Zn, Co, Cr and Fe) and macroelements (Ca, Mg, Na and K) in blood of laying hens fed on the diet enriched with marine macroalgae were higher than in the control group [43]. Sodium concentration in blood serum of laying hens decreased when they ate feed containing 2% of *Sarcodiotheca gaudichaudii* and *Chondrus crispus* when compared to the control [31]. The same authors also found that there was a non–significant impact of red seaweeds on the content of phosphorus, chlorine, calcium, potassium, total protein, glucose, creatine kinase, aspartate amino transferase and uric acid in blood serum.

9. Advantages and Disadvantages of Seaweeds in Poultry Nutrition and Future Prospects

When using seaweeds in poultry nutrition, their advantages and disadvantages should be taken into account (Table 3).

Table 3. Examples of advantages and disadvantages of seaweeds in poultry feeding.

ADVANTAGES	DISADVANTAGES

■ Effect on egg composition:
  – Increase in the n–3 fatty acids content in eggs [44]
  – Increase in the n–9 fatty acids content in eggs [70]
  – Improvement of the content of protein in egg [56]
  – Decrease in egg cholesterol and triglycerides [40,44,46,47,57,70,80]
■ Improvement of meat quality:
  – Lower fat content [20,49,53,83,87,88]
  – Higher content of protein [83]
■ Egg-laying rate and egg quality parameters:
  – Increase in laying rate [31,35,40,43,80]
  – Increase in the egg weight [40,43,46,51,56,81]
  – Increase in the yolk color [40,43,44,46,47,55–58,65,70,81]
  – Increase in the albumen height [44,50,56]
  – Increase in the Haugh unit [44,51,56,81]
  – Increase in the shell thickness [40,43,46,51,56,81]
  – Increase in egg–shape index [40,51,80]
  – Increase in eggshell strength [47]
■ Effect on blood profile:
  – Decrease in plasma cholesterol, LDL, VLDL and triglycerides [38,40,48,70,80]
  – Increase in enzymatic antioxidant activity [40,80]
  – Increase in lymphocyte number [36]
  – Decrease in sodium concentration [31]
  – Improvement of heterophils to lymphocytes ratio [80]
■ Improvement of growth performance:
  – Increase in body weight [20,30,32,43,62,63]
  – Increase in body weight gain [18,36]
  – Increase in feed intake [20,48,62]
  – Improving feed conversion ratio [20,30,32,48,63]
  – Decrease in mortality rate [32,48]
■ Improvement in fertility and hatchability:
  – Increase in fertility [40,89,90]
  – Increase in hatchability [40,89]
■ Can act as prebiotics [28,31,34,40,48]
■ Enhancement of immune functions and intestinal villi [18,32,34,35,37,48,80]
■ Boosting useful bacteria [31,34,36,93,94]
	■ Seaweeds polysaccharides can bind nutrients and inhibit their absorption in the gastrointestinal tract [60]
	■ Seaweeds antinutrients can interfere with digestion and feed utilization processes [67,68,98]
	■ Seaweeds can contain toxic metals [42,60,67,68]
	■ High salt content in seaweeds can cause diarrhea and poultry death [59,72]
	■ Seasonal and geographical variations in chemical composition of seaweeds [19,20,42,64,67]
	■ Cultivation and processing methods and costs may impact the nutritional value of seaweeds [99–101]
	■ The high foot pad dermatitis score was found in broilers fed algae meal [96]

On the basis of the literature reviewed in the present article, it could be concluded that seaweeds at different levels and forms have a positive influence on the growth and performance of poultry as well as their blood profile and health. When seaweeds are added to the feed in the right proportions, at low inclusion levels, they also enrich poultry products (meat, eggs) with biologically active compounds. The use of algae as natural pigments can be of great value for the poultry industry since they can increase yolk color without resorting to synthetic carotenoids. The economic parameters of eggs such as weight, shell weight, thickness and strength are especially important to producers and consumers. As it was shown in the present paper, these parameters can be improved by seaweeds which should have proper chemical composition, lack of toxic metals or compounds that can act as antinutritional factors.

Economic aspects may limit the use of seaweeds in poultry diet. Studies on this problem are scarce. Currently, seaweeds are not grown on a meaningful scale. According to Burg et al. (2012), the total seaweed production expenses, without collection and transportation, are approximately €1,000 and €1,500 for each ton of dry matter [101]. Therefore, seaweeds naturally occurring in the environment have the greatest potential to be used as feed additives. The application of seaweeds in the sun–dried and ground form is the most beneficial. It was shown that seaweed processing had no significant effect on poultry growth or performance. Each seaweed processing is treated as an added cost. Findings of Zeweil et al. (2019) showed that the Japanese quail fed on diets supplemented with green and brown seaweeds (1.5% and 3%, respectively) generated the greatest net income and relative efficacy (102.5 and 104.2 for green seaweeds and 107.09 and 101.11 for brown seaweeds at the dose of 1.5% and 3%, consecutively) in comparison with the control group [40]. Seaweeds have the potential to be commonly used as feed additives not only thanks to their properties, but also due to the fact that the search for new, cheaper, safe feed additives is a priority of the poultry industry.

Author Contributions: Conceptualization, I.M. and K.M.; writing—original draft preparation, I.M. and K.M.; writing, review and editing, I.M. and K.M. All authors have read and agreed to the published version of the manuscript.

Funding: I.M. was funded by a grant entitled: "The effect of bioactive algae enriched by biosorption in the certain minerals such as Cr(III), Mg(II) and Mn(II) on the status of glucose in the course of metabolic syndrome horses. Evaluation in vitro and *in vivo*" (No 2015/18/E/NZ9/00607) from the National Science Center in Poland (part about seaweeds, biologically active compounds, their forms in animal nutrition).

Conflicts of Interest: The authors declare no conflict of interest.

References

1. Ahaotu, E.O.; De los Ríos, P.; Ibe, L.C.; Singh, R.R. Climate change in poultry production system—A review. *Acta Sci. Agric.* **2019**, *3*, 113–117.
2. United States Department of Agriculture (USDA). Livestock and Poultry: World Markets and Trade. Foreign Agricultural Service, USA. 2020. Available online: https://apps.fas.usda.gov/psdonline/circulars/livestock_poultry.pdf (accessed on 27 July 2020).
3. El-Hack, M.E.A.; Mahrose, K.M.; Askar, A.A.; Alagawany, M.; Arif, M.; Saeed, M.; Abbasi, F.; Soomro, R.N.; Siyal, F.A.; Chaudhry, M.T. Single and combined impacts of vitamin A and selenium in diet on productive performance, egg quality, and some blood parameters of laying hens during hot season. *Biol. Trace Element Res.* **2016**, *177*, 169–179. [CrossRef] [PubMed]
4. Saeed, M.; El-Hack, M.E.A.; Arif, M.; El-Hindawy, M.M.; Attia, A.I.; Mahrose, K.M.; Bashir, I.; Siyal, F.A.; Arain, M.A.; Fazlani, S.A.; et al. Impacts of distiller's dried grains with solubles as replacement of soybean meal plus vitamin E supplementation on production, egg quality and blood chemistry of laying hens. *Ann. Anim. Sci.* **2017**, *17*, 849–862. [CrossRef]
5. Nordhagen, S.; Klemm, R. Implementing small-scale poultry-for-nutrition projects: Successes and lessons learned. *Matern. Child Nutr.* **2018**, *14*, e12676. [CrossRef] [PubMed]
6. Daghir, N.J. *Poultry Production in Hot Climates*, 6th ed.; Cromwell Press: Trowbridge, UK, 1995; pp. 1–12.
7. Alagawany, M.; Mahrose, K.M. Influence of different levels of certain essential amino acids on the performance, egg quality criteria and economics of Lohmann Brown laying hens. *Asian J. Poult. Sci.* **2014**, *8*, 82–96. [CrossRef]
8. Farghly, M.F.A.; Mahrose, K.M.; Galal, A.E.; Ali, R.M.; Ahmad, E.A.M.; Rehman, Z.U.; Ullah, Z.; Ding, C. Implementation of different feed withdrawal times and water temperatures in managing turkeys during heat stress. *Poult. Sci.* **2018**, *97*, 3076–3084. [CrossRef]
9. Farghly, M.F.A.; Mahrose, K.M.; Cooper, R.; Ullah, Z.; Rehman, Z.U.; Ding, C. Sustainable floor type for managing turkey production in a hot climate. *Poult. Sci.* **2018**, *97*, 3884–3890. [CrossRef]
10. El-Hack, M.E.A.; Mahrose, K.M.; Attia, F.A.M.; Swelum, A.A.; Taha, A.E.; Shewita, R.; Hussein, E.-S.O.S.; Alowaimer, A.N. Laying performance, physical, and internal egg quality criteria of hens fed distillers dried grains with solubles and exogenous enzyme mixture. *Animals* **2019**, *9*, 150. [CrossRef]
11. Abou-Kassem, D.E.; Ashour, E.A.; Alagawany, M.; Mahrose, K.M.; Rehman, Z.U.; Ding, C. Effect of feed form and dietary protein level on growth performance and carcass characteristics of growing geese. *Poult. Sci.* **2019**, *98*, 761–770. [CrossRef]
12. Mahrose, K.M.; El-Hack, M.E.A.; Mahgoub, S.A.; Attia, F.A.M. Influences of stocking density and dietary probiotic supplementation on growing Japanese quail performance. *An. Acad. Bras. Cienc.* **2019**, *91*, e20180616. [CrossRef]

13. Mahrose, K.M.; El-Hack, M.E.A.; Amer, S.A. Influences of dietary crude protein and stocking density on growth performance and body measurements of ostrich chicks. *An. Acad. Bras. Cienc.* **2019**, *91*, e20180479. [CrossRef] [PubMed]

14. Rizk, Y.S.; Fahim, H.N.; Beshara, M.M.; Mahrose, K.M.; Awad, A.L. Response of duck breeders to dietary L-Carnitine supplementation during summer season. *An. Acad. Bras. Cienc.* **2019**, *91*, e20180907. [CrossRef] [PubMed]

15. Cabrita, A.R.J.; Maia, M.; Oliveira, H.M.; Pinto, I.S.; Almeida, A.; Pinto, E.; Fonseca, A.J.M. Tracing seaweeds as mineral sources for farm-animals. *J. Appl. Phycol.* **2016**, *28*, 3135–3150. [CrossRef]

16. Holdt, S.L.; Kraan, S. Bioactive compounds in seaweed: Functional food applications and legislation. *J. Appl. Phycol.* **2011**, *23*, 543–597. [CrossRef]

17. Michalak, I.; Chojnacka, K. Algae as production systems of bioactive compounds. *Eng. Life Sci.* **2015**, *15*, 160–176. [CrossRef]

18. Qadri, S.S.N.; Biswas, A.; Mandal, A.B.; Kumawat, M.; Saxena, R.; Nasir, A.M. Production performance, immune response and carcass traits of broiler chickens fed diet incorporated with *Kappaphycus Alvarezii*. *J. Appl. Phycol.* **2018**, *31*, 753–760. [CrossRef]

19. Øverland, M.; Mydland, L.T.; Skrede, A. Marine macroalgae as sources of protein and bioactive compounds in feed for monogastric animals. *J. Sci. Food Agric.* **2018**, *99*, 13–24. [CrossRef]

20. Erum, T.; Frias, G.G.; Cocal, C.J. *Sargassum muticum* as feed substitute for broiler. *Asia Pacific. J. Educ. Arts Sci.* **2017**, *4*, 6–9.

21. Evans, F.; Critchley, A.T. Seaweeds for animal production use. *J. Appl. Phycol.* **2014**, *26*, 891–899. [CrossRef]

22. Makkar, H.; Tran, G.; Heuzé, V.; Giger-Reverdin, S.; Lessire, M.; LeBas, F.; Ankers, P. Seaweeds for livestock diets: A review. *Anim. Feed Sci. Technol.* **2016**, *212*, 1–17. [CrossRef]

23. Corino, C.; Modina, S.; Di Giancamillo, A.; Chiapparini, S.; Rossi, R. Seaweeds in pig nutrition. *Animals* **2019**, *9*, 1126. [CrossRef] [PubMed]

24. Michalak, I.; Marycz, K. Algae as a promising feed additive for horses. In *Seaweeds as Plant Fertilizer, Agricultural Biostimulants and Animal Fodder*; Pereira, L., Bahcevandziev, K., Joshi, N.H., Eds.; CRC Press, Taylor & Francis Group: Boca Raton, FL, USA, 2019; Volume 7, pp. 128–142.

25. Morais, T.; Inácio, A.; Coutinho, T.; Ministro, M.; Cotas, J.; Pereira, L.; Bahcevandziev, K. Seaweed potential in the animal feed: A review. *J. Mar. Sci. Eng.* **2020**, *8*, 559. [CrossRef]

26. Haberecht, S.; Wilkinson, S.; Roberts, J.; Wu, S.-B.; Swick, R. Unlocking the potential health and growth benefits of macroscopic algae for poultry. *World's Poult. Sci. J.* **2018**, *74*, 5–20. [CrossRef]

27. Kulshreshtha, G.; Hincke, M.T.; Prithiviraj, B.; Critchley, A.T. A Review of the varied uses of macroalgae as dietary supplements in selected poultry with special reference to laying hen and broiler chickens. *J. Mar. Sci. Eng.* **2020**, *8*, 536. [CrossRef]

28. Yan, G.L.; Guo, Y.M.; Yuan, J.M.; Liu, D.; Zhang, B.K. Sodium alginate oligosaccharides from brown algae inhibit *Salmonella* Enteritidis colonization in broiler chickens. *Poult. Sci.* **2011**, *90*, 1441–1448. [CrossRef] [PubMed]

29. Wiseman, M. *Evaluation of Tasco® as a Candidate Prebiotic in Broiler Chickens*; Dalhousie University: Halifax, NS, Canada, 2012; Available online: https://dalspace.library.dal.ca/bitstream/handle/10222/14443/Wiseman_Melissa_MSc.__Animal_Science_February_2012.pdf?sequence=3&isAllowed=y (accessed on 25 April 2020).

30. Islam, M.M.; Ahmed, S.T.; Mun, H.S.; Kim, Y.J.; Yang, C.J. Effect of Sea Tangle (*Laminaria japonica*) and charcoal supplementation as alternatives to antibiotics on growth performance and meat quality of ducks. *Asian-Australas. J. Anim. Sci.* **2014**, *27*, 217–224. [CrossRef]

31. Kulshreshtha, G.; Rathgeber, B.; Stratton, G.; Thomas, N.; Evans, F.; Critchley, A.T.; Hafting, J.; Prithiviraj, B. Feed supplementation with red seaweeds, *Chondrus crispus* and *Sarcodiotheca gaudichaudii*, affects performance, egg quality, and gut microbiota of layer hens. *Poult. Sci.* **2014**, *93*, 2991–3001. [CrossRef]

32. Choi, Y.J.; Lee, S.R.; Oh, J.-W. Effects of dietary fermented seaweed and seaweed fusiforme on growth performance, carcass parameters and immunoglobulin concentration in broiler chicks. *Asian-Australas. J. Anim. Sci.* **2014**, *27*, 862–870. [CrossRef]

33. Karimi, S.H. *Effects of Red Seaweed (Palmaria Palmata) Supplemented Diets Fed to Broiler Chickens Raised under Normal or Stressed Conditions*; Dalhousie University: Halifax, NS, Canada, 2015; Available online: https://dalspace.library.dal.ca/bitstream/handle/10222/64662/Karimi--Seyed_Hossein--MSc--September_21.pdf?sequence=3&isAllowed=y (accessed on 24 April 2020).

34. Kulshreshtha, G.; Rathgeber, B.; MacIsaac, J.; Boulianne, M.; Brigitte, L.; Stratton, G.; Thomas, N.A.; Critchley, A.T.; Hafting, J.; Prithiviraj, B. Feed supplementation with red seaweeds, *Chondrus crispus* and *Sarcodiotheca gaudichaudii*, reduce *Salmonella* Enteritidis in laying hens. *Front. Microbiol.* **2017**, *8*, 567. [CrossRef]

35. Choi, Y.; Lee, E.; Na, Y.; Lee, S. Effects of dietary supplementation with fermented and non-fermented brown algae by-products on laying performance, egg quality, and blood profile in laying hens. *Asian-Australasian J. Anim. Sci.* **2018**, *31*, 1654–1659. [CrossRef]

36. Bai, J.; Wang, R.; Yan, L.; Feng, J. Co-Supplementation of dietary seaweed powder and antibacterial peptides improves broiler growth performance and immune function. *Braz. J. Poult. Sci.* **2019**, *21*, 1–9. [CrossRef]

37. El-Deekx, A.; Bri, A.M.; Brikaa, A.M. Effect of different levels of seaweed in starter and finisher diets in pellet and mash form on performance and carcass quality of ducks. *Int. J. Poult. Sci.* **2009**, *8*, 1014–1021. [CrossRef]

38. Frasiska, N.; Suprijatna, E.; Susanti, S. Effect of diet containing *Gracilaria* sp. waste and multi-enzyme additives on blood lipid profile of local duck. *Anim. Prod.* **2016**, *18*, 22. [CrossRef]

39. Karu, P.; Selvan, S.; Gopi, H.; Manobhavan, M. Effect of macroalgae supplementation on growth performance of Japanese quails. *Int. J. Curr. Microbiol. Appl. Sci.* **2018**, *7*, 1039–1041. [CrossRef]

40. Zeweil, S.H.; Abu Hafsa, S.H.; Zahran, S.M.; Ahmed, M.S.; Abdel–Rahman, N. Effects of dietary supplementation with green and brown seaweeds on laying performance, egg quality, and blood lipid profile and antioxidant capacity in laying Japanese quail. *Egypt. Poult. Sci. J.* **2019**, *39*, 41–59. [CrossRef]

41. Ventura, M.; Castañon, J.; McNab, J. Nutritional value of seaweed (*Ulva rigida*) for poultry. *Anim. Feed. Sci. Technol.* **1994**, *49*, 87–92. [CrossRef]

42. Michalak, I.; Chojnacka, K. Multielemental analysis of macroalgae from the Baltic Sea by ICP-OES to monitor environmental pollution and assess their potential uses. *Int. J. Environ. Anal. Chem.* **2009**, *89*, 583–596. [CrossRef]

43. Michalak, I.; Chojnacka, K.; Dobrzański, Z.; Górecki, H.; Zielińska, A.; Korczyński, M.; Opaliński, S. Effect of macroalgae enriched with microelements on egg quality parameters and mineral content of eggs, eggshell, blood, feathers and droppings. *J. Anim. Physiol. Anim. Nutr.* **2010**, *95*, 374–387. [CrossRef]

44. Carrillo-Dominguez, S.; López, E.; Casas, M.M.; Avila, E.; Castillo, R.M.; Carranco, M.E.; Calvo, C.; Pérez-Gil, F. Potential use of seaweeds in the laying hen ration to improve the quality of n-3 fatty acid enriched eggs. *J. Appl. Phycol.* **2008**, *20*, 721–728. [CrossRef]

45. Carrillo-Dominguez, S.; Ríos, V.H.; Calvo, C.; Carranco, M.E.; Casas, M.; Pérez-Gil, F. n-3 fatty acid content in eggs laid by hens fed with marine algae and sardine oil and stored at different times and temperatures. *J. Appl. Phycol.* **2012**, *24*, 593–599. [CrossRef]

46. Wang, S.; Hui, J.Y.; Hua, W.L.; Hua, Z.F.; Ting, L.Y. *Enteromorpha prolifera* supplemental level: Effects on laying performance, egg quality, immune function and microflora in feces of laying hens. *Chin. J. Anim. Nutr.* **2013**, *25*, 1346–1352.

47. Li, Q.; Luo, J.; Wang, C.; Tai, W.; Wang, H.; Zhang, X.; Liu, K.; Jia, Y.; Lyv, X.; Wang, L.; et al. Ulvan extracted from green seaweeds as new natural additives in diets for laying hens. *J. Appl. Phycol.* **2018**, *30*, 2017–2027. [CrossRef]

48. Cañedo-Castro, B.; Piñón-Gimate, A.; Carrillo, S.; Ramos, D.; Casas-Valdez, M. Prebiotic effect of *Ulva rigida* meal on the intestinal integrity and serum cholesterol and triglyceride content in broilers. *J. Appl. Phycol.* **2019**, *31*, 3265–3273. [CrossRef]

49. Abudabos, A.M.; Okab, A.B.; Aljumaah, R.; Samara, E.; Abdoun, K.A.; Al-Haidary, A.A. Nutritional value of green seaweed (*Ulva lactuca*) for broiler chickens. *Ital. J. Anim. Sci.* **2013**, *12*, 28. [CrossRef]

50. Stupart, C.M. *Supplementation of Red Seaweed (Chondrus crispus) and Tasco® (Ascophyllum nodosum) in Laying Hen Diets*; Dalhousie University: Halifax, NS, Canada, 2019; Available online: https://dalspace.library.dal.ca/bitstream/handle/10222/76824/Stupart--Cassandra--MSc--AGRI--December--2019.pdf?sequence=5&isAllowed=y (accessed on 25 April 2020).

51. Mandal, A.B.; Biswas, A.; Mir, N.A.; Tyagi, P.K.; Kapil, D.; Biswas, A.K. Effects of dietary supplementation of *Kappaphycus alvarezii* on productive performance and egg quality traits of laying hens. *J. Appl. Phycol.* **2019**, *31*, 2065–2072. [CrossRef]

52. El-Deek, A.; Brikaa, M.A. Nutritional and biological evaluation of marine seaweed as a feedstuff and as a pellet binder in poultry diet. *Int. J. Poult. Sci.* **2009**, *8*, 875–881. [CrossRef]

53. El-Naga, M.A.; Megahed, M. Impact of brown algae supplementation in drinking water on growth performance and intestine histological changes of broiler chicks. *Egypt. J. Nutr. Feed.* **2018**, *21*, 495–507. [CrossRef]

54. Bonos, E.; Kargopoulos, A.; Nikolakakis, I.; Florou-Paneri, P.; Christaki, E. The seaweed *Ascophyllum nodosum* as a potential functional ingredient in chicken nutrition. *J. Oceanogr. Mar. Res.* **2017**, *4*, 140. [CrossRef]

55. Strand, A.; Herstad, O.; Liaaen-Jensen, S. Fucoxanthin metabolites in egg yolks of laying hens. *Comp. Biochem. Physiol. Part A Mol. Integr. Physiol.* **1998**, *119*, 963–974. [CrossRef]

56. Rendón, U.; Carrillo, S.; Arellano, L.G.; Casas, M.M.; Pérez, F.; Avila, E. Chemical composition of the residue of alginates (*Macrocystis pyrifera*) extraction. Its utilization in laying hens feeding. *Cuban J. Agricult. Sci.* **2003**, *37*, 287–293.

57. Carrillo, S.; Bahena, A.; Casas, M.; Carranco, M.E.; Calvo, C.C.; Ávila, E.; Pérez–Gi, F. The alga *Sargassum* spp. as alternative to reduce egg cholesterol content. *Cuban J. Agricult. Sci.* **2012**, *46*, 181–186.

58. El–Deek, A.A.; Al–Harthi, M.A. Nutritive value of treated brown marine algae in pullet and laying diets. World Poultry Science Association. In Proceedings of the 19th European Symposium on Quality of Poultry Meat, 13th European Symposium on the Quality of Eggs and Egg Products, Turku, Finland, 21–25 June 2009; pp. 1–12.

59. Dewi, Y.L.; Yuniza, A.; Nuraini; Sayuti, K.; Mahata, M.E. Immersion of *Sargassum binderi* seaweed in river water flow to lower salt content before use as feed for laying hens. *Int. J. Poult. Sci.* **2017**, *17*, 22–27. [CrossRef]

60. Dewi, Y.L.; Yuniza, A.; Sayuti, K.; Mahata, M.E.; Nuraini, N. Fermentation of *Sargassum binderi* seaweed for lowering alginate content of feed in laying hens. *J. World's Poult. Res.* **2019**, *9*, 147–153. [CrossRef]

61. Al–Harthi, M.A.; El–Deek, A.A. Nutrient profiles of brown marine algae (*Sargassum dentifebium*) as affected by different processing methods for chickens. *J. Food Agricult. Environ.* **2012**, *10*, 475–480.

62. El–Deek, A.A.; Al–Harthi, M.A.; Abdalla, A.A.; Elbanoby, M.M. The use of brown algae meal in finisher broiler diets. *Egypt. Poult. Sci.* **2011**, *31*, 767–781.

63. Athis Kumar, K. Effect of *Sargassum wightii* on growth, carcass and serum qualities of broiler chickens. *Vet. Sci. Res.* **2018**, *3*, 000156.

64. Fleurence, J. Seaweed proteins: Biochemical, nutritional aspects and potential uses. *Trends Food Sci. Technol.* **1999**, *10*, 25–28. [CrossRef]

65. Herber-McNeill, S.M.; Van Elswyk, M.E. Dietary marine algae maintains egg consumer acceptability while enhancing yolk color. *Poult. Sci.* **1998**, *77*, 493–496. [CrossRef]

66. González-Esquerra, R.; Leeson, S. Alternatives for enrichment of eggs and chicken meat with omega-3 fatty acids. *Can. J. Anim. Sci.* **2001**, *81*, 295–305. [CrossRef]

67. Stengel, D.B.; Connan, S.; Popper, Z.A. Algal chemodiversity and bioactivity: Sources of natural variability and implications for commercial application. *Biotechnol. Adv.* **2011**, *29*, 483–501. [CrossRef]

68. Rajauria, G. Seaweeds: A sustainable feed source for livestock and aquaculture. *Seaweed Sustain.* **2015**, 389–420. [CrossRef]

69. Lahaye, M.; Vigouroux, J. Liquefaction of dulse (*Palmaria palmata* (L.) Kuntze) by a commercial enzyme preparation and purified endo–β–1,4–D–xylanase. *J. Appl. Phycol.* **1992**, *4*, 329–337. [CrossRef]

70. Al-Harthi, M.A.; El–Deek, A.A. Effect of different dietary concentrations of brown marine algae (*Sargassum dentifebium*) prepared by different methods on plasma and yolk lipid profiles, yolk total carotene and lutein plus zeaxanthin of laying hens. *Ital. J. Anim. Sci.* **2012**, *11*, 347–353. [CrossRef]

71. Yuan, Y.V. *Marine Algal Constituents*; Barrow, C., Shahidi, F., Eds.; Marine Nutraceuticals and Functional Foods; CRC: Boca Raton, FL, USA, 2008; pp. 259–296.

72. Zahid, P.B.; Aisha, K.; Ali, A. Green seaweed as component of poultry feed. *Bangladesh J. Bot.* **1995**, *24*, 153–156.

73. Charoensiddhi, S.; Abraham, R.E.; Su, P.; Zhang, W. Seaweed and seaweed-derived metabolites as prebiotics. *Adv. Food Nutr. Res.* **2020**, *91*, 97–156. [CrossRef]

74. Gibson, G.R.; Beatty, E.R.; Wang, X.; Cummings, J.H. Selective stimulation of bifidobacteria in the human colon by oligofructose and inulin. *Gastroenterology* **1995**, *108*, 975–982. [CrossRef]

75. Sun, J.; Song, H.L.; Zhao, J.; Xiao, Y.; Qi, R.; Lin, Y.T. Effects of different dietary levels of *Enteromorpha prolifera* on nutrient availability and digestive enzyme activities of broiler chickens. *Chin. J. Anim. Nutr.* **2010**, *22*, 1658–1664.

76. Herber, S.M.; Van Elswyk, M.E. Dietary marine algae promotes efficient deposition of n-3 fatty acids for the production of enriched shell eggs. *Poult. Sci.* **1996**, *75*, 1501–1507. [CrossRef]

77. Simopoulos, A.P. Human requirement for n-3 polyunsaturated fatty acids. *Poult. Sci.* **2000**, *79*, 961–970. [CrossRef]

78. FAO. *Fats and Fatty Acids in Human Nutrition. Report of an Expert Consultation*; FAO Food and Nutrition: Geneva, Switzerland, 2008.

79. Qi, H.; Sheng, J. The antihyperlipidemic mechanism of high sulfate content ulvan in rats. *Mar. Drugs* **2015**, *13*, 3407–3421. [CrossRef]

80. Rizk, Y.S.; Ismail, I.I.; Abu Hafsa, S.H.; Eshera, A.A.; Tawfeek, F.A. Effect of dietary green tea and dried seaweed on productive and physiological performance of laying hens during late phase of production. *Egypt. Poult. Sci.* **2017**, *37*, 685–706. [CrossRef]

81. Al-Harthi, M.A.; El-Deek, A.A. The effects of preparing methods and enzyme supplementation on the utilization of brown marine algae (*Sargassum dentifebium*) meal in the diet of laying hens. *Ital. J. Anim. Sci.* **2011**, *10*, 48. [CrossRef]

82. Bradbury, E.J.; Wilkinson, S.J.; Cronin, G.M.; Walk, C.L.; Cowieson, A.J. The effect of marine calcium source on broiler leg integrity. In Proceedings of the 23rd Annual Australian Poultry Science Symposium, Sydney, Australia, 19–22 February 2012; pp. 85–88.

83. Zahid, P.B.; Ali, A.; Zahid, M.J. Brown seaweeds as supplement for broiler feed. *Hamdard Med.* **2001**, *44*, 98–101.

84. Koh, T.S.; Im, J.T.; Park, I.K.; Lee, H.J.; Choi, D.Y.; Choi, C.J.; Lee, H.G.; Choi, Y.J. Effect of dietary brown seaweed levels on the protein and energy metabolism in broiler chicks activated acute phase response. *J. Anim. Sci. Technol. (Kor.)* **2005**, *47*, 379–390.

85. Ali, A.; Memon, M.S. Incorporation of *Enteromorpha procera* Ahlner as nutrition supplement in chick's feed. *Int. J. Biol. Biotechnol.* **2008**, *5*, 211–214.

86. Aisha, K.; Zahid, P.B. Brown seaweeds as supplementary feed for poultry. *Int. J. Phycol. Phycochem.* **2009**, *5*, 17–20.

87. Wang, S.; Shi, X.; Zhou, C.; Lin, Y. *Entermorpha prolifera*: Effects on performance, carcass quality and small intestinal digestive enzyme activities of broilers. *Chin. J. Anim. Nutr.* **2013**, *25*, 1332–1337.

88. Martínez, Y.; Ayala, L.; Hurtado, C.; Más, D.; Rodríguez, R. Effects of dietary supplementation with red algae powder (*Chondrus crispus*) on growth performance, carcass traits, lymphoid organ weights and intestinal pH in broilers. *Braz. J. Poult. Sci.* **2019**, *21*, 1–7.

89. Bratova, K.; Ganovski, K. Effect of Black Sea algae on chicken egg production and on chick embryo development. *Vet. Med. Nauki.* **1982**, *19*, 99–105.

90. Vidanarachchi, J.; Mikkelsen, L.L.; Sims, I.; Iji, P.A.; Choct, M. Phytobiotics: Alternatives to antibiotic growth promoters in monogastric animal feeds. *Rec. Adv. Anim. Nutr. Aust.* **2005**, *15*, 131–144.

91. Ahmadi, A.; Moghadamtousi, S.Z.; Abubakar, S.; Zandi, K. Antiviral potential of algae polysaccharides isolated from marine sources: A review. *BioMed Res. Int.* **2015**, *2015*, 1–10. [CrossRef] [PubMed]

92. Elizondo-Gonzalez, R.; Cruz-Suárez, L.E.; Marie, D.R.; Mendoza-Gamboa, E.; Rodriguez-Padilla, C.; Trejo-Ávila, L.M. *In vitro* characterization of the antiviral activity of fucoidan from *Cladosiphon okamuranus* against Newcastle Disease Virus. *Virol. J.* **2012**, *9*, 307. [CrossRef] [PubMed]

93. Siahaan, E.A.; Pendleton, P.; Woo, H.-C.; Chun, B.-S. Brown seaweed (*Saccharina japonica*) as an edible natural delivery matrix for allyl isothiocyanate inhibiting food-borne bacteria. *Food Chem.* **2014**, *152*, 11–17. [CrossRef]

94. Radulovich, R.; Umanzor, S.; Cabrera, R.; Mata, R. Tropical seaweeds for human food, their cultivation and its effect on biodiversity enrichment. *Aquaculture* **2015**, *436*, 40–46. [CrossRef]

95. Leonard, S.G.; Sweeney, T.; Bahar, B.; Lynch, B.P.; O'Doherty, J.V. Effect of maternal fish oil and seaweed extract supplementation on colostrum and milk composition, humoral immune response, and performance of suckled piglets. *J. Anim. Sci.* **2010**, *88*, 2988–2997. [CrossRef]

96. El-Wahab, A.A.; Visscher, C.; Kamphues, J. Impact of different dietary protein sources on performance, litter quality and foot pad dermatitis in broilers. *J. Anim. Feed. Sci.* **2018**, *27*, 148–154. [CrossRef]

97. Zhang, Q.; Li, N.; Liu, X.; Zhao, Z.; Li, Z.; Xu, Z. The structure of a sulfated galactan from *Porphyra haitanensis* and its *in vivo* antioxidant activity. *Carbohydr. Res.* **2004**, *339*, 105–111. [CrossRef]

98. Francis, G.; Makkar, H.P.; Becker, K. Antinutritional factors present in plant-derived alternate fish feed ingredients and their effects in fish. *Aquaculture* **2001**, *199*, 197–227. [CrossRef]

99. Huyghebaert, G.; Ducatelle, R.; Van Immerseel, F. An update on alternatives to antimicrobial growth promoters for broilers. *Veter. J.* **2011**, *187*, 182–188. [CrossRef]

100. Buschmann, A.H.; Camus, C.; Infante, J.; Neori, A.; Israel, A.; Hernández-González, M.C.; Pereda, S.V.; Pinchetti, J.L.G.; Golberg, A.; Tadmor-Shalev, N.; et al. Seaweed production: Overview of the global state of exploitation, farming and emerging research activity. *Eur. J. Phycol.* **2017**, *52*, 391–406. [CrossRef]

101. van den Burg, S.; Stuiver, M.; Veenstra, F.; Bikker, P.; López Contreras, A.; Palstra, A.; Broeze, J.; Jansen, H.; Jak, R.; Gerritsen, A.; et al. *A Triple P Review of the Feasibility of Sustainable Offshore Seaweed Production in the North. Sea*; Wageningen UR (University & Research Centre): Wageningen, The Netherlands, 2012.

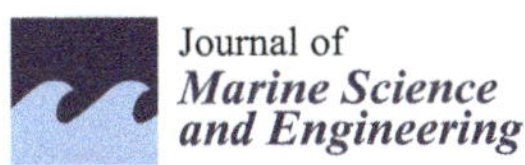

Journal of
Marine Science and Engineering

Review

The Animal Kingdom, Agriculture··· and Seaweeds

Melania L. Cornish [1,*], Michéal Mac Monagail [2] and Alan T. Critchley [3]

[1] Acadian Seaplants Limited, Cornwallis Park, NS B0S1A0, Canada
[2] Arramara Teoranta, Oyster Bay, Kilkieran, Co., H91 HD86 Galway, Ireland; mmacmonagail@arramara.ie
[3] Verschuren Centre for Sustainability in Energy and the Environment, Cape Breton University, Sydney, NS B1P 6L2, Canada; alan.critchley2016@gmail.com
* Correspondence: lcornish@acadian.ca

Received: 27 June 2020; Accepted: 28 July 2020; Published: 30 July 2020

Abstract: Marine macroalgae (seaweeds), are amongst the first multicellular organisms and, as such, the precursors to land plants. By the time 'land' animals arrived on the scene, terrestrial plants were plentiful and varied, and herbivorous diets developed in concert with the food sources most commonly available. However, skip forward several hundred millennia, and with the advent of agriculture, approximately 10,000 years ago, dietary diversity began to change. Today, the world is experiencing increasingly higher rates of debilitating, non-communicable diseases—might there be a connection? This paper reviews scientific evidence for the judicious use of various seaweeds in the reduction of heat stress, enhanced immunity, improved growth performance, and methane reduction in animals. The extensive, (super) prebiotic effects of selected macroalgae will also be highlighted. Key studies conducted across the animal kingdom provide considerable support that there is an overwhelming need for the guided and wise applications of increased usage of selected seaweeds in feed, food and supplements. Particular attention will be paid to the bioactive components, and nutraceutical qualities, of various seaweeds, i.e., the brown, *Saccharina* (*Laminaria*) spp. and *Ascophyllum nodosum*, and the red alga *Chondrus crispus*. Suggestions are put forward for benefits to be derived from their further applications.

Keywords: macroalgae (seaweeds); feed; food; *Homo sapiens*; agriculture; health

---

1. Introduction

With the lack of lignified tissues and thereby extensive fossilized evidence, it is challenging for researchers to prove definitively that seaweeds were eaten as a crucial part of the diet by early animals, including *Homo sapiens*. However, some of the tools most commonly used by scientists in their attempts to determine the components of ancient diets include reconstructions of the biomechanics of fossilized jaws, bone and teeth isotopic data, and tooth wear patterns [1]. Isotopic analysis has also been carried out on numerous fossilized remains of early hominids, the results of which indicated that before 4 mya, most hominid diets consisted primarily of C_3 plants (trees, fruits, shrubs, and non-grassy herbs and forbs), akin to the diets of non-human primates. By about 3.5 mya, multiple taxa began to increasingly incorporate C_4 foods (primarily grasses and sedges) into their diets, although the trend and ratio varied by region [2,3]. Even this information though, is problematic in determining if early humans and other animals consumed seaweeds, as the ^{13}C-^{12}C ratio varies widely amongst macroalgal species. Maberly et al., 1992 analyzed no less than 9 species of green, 15 brown, and 22 red seaweeds collected from various places around the east coast of Scotland. Their results ranged from 8.81 to 34.74%, whereas C_4 plants are usually around 12% and C_3 plants around 28% [4,5], and this effectively discounts isotopic analysis at this time, as a tool to define early seaweed consumption.

This leaves only, so far, tooth wear potentially from sand particles, and the knowledge that a stable supply of all the essential nutrients for neonatal brain growth has a high probability of exerting dietary influence on development. This necessitates a specific period beginning with the maternal diet prior to, during, and after a lengthy gestation, followed by months, or years of nursing. An example of some of these nutrients includes polyunsaturated fatty acids (PUFAs), particularly docosahexaenoic acid (DHA:C22:6, n-3) and arachidonic acid (AA:C20:4, n-6), and critical non-residual nutrients such as zinc, iodine, and vitamin B_{12} [6]. All are available to foragers in coastal marine environments around the world, and seaweeds, which have been growing along the worlds' shorelines for eons and were present well before the animals moved from the sea onto the land.

The most significant changes in human brain development occurred over the past 2.5–2 million years [7–9]. Modern-day humans now boast ownership of a precious and complex organ that functions as the epicenter of human physical existence, intelligence, and the source of all those features that define humanity. Of utmost importance to healthy neonatal brain growth and development, is the quality of the maternal diet prior to, during, and after the lengthy gestation period typical of humans. To accommodate the nutritional necessities for enhancing brain size and the associated cognitive abilities over the evolutionary long-term, a diet containing all the nutrients is likely to be a necessity over multiple generations. Family units living and eating in coastal environments 2.5–2 million years ago would have the best chances of reaching and maintaining nutritional integrity and the associated enhanced cognitive abilities. Albeit this is not proof, and absolute proof remains elusive, if not impossible, but a logical postulation.

Ironically, there is evidence today that the human brain is now actually shrinking in size. While speculation as to the reasons and implications is varied, one of the most common suggestions is that of worsening nutrition. In reality, there are most likely many factors involved and the reasons for it are quite complicated, but over the past 20,000 years, the average volume of the human male brain has decreased from 1500 cm^3 to 1350 cm^3 [10,11]. It is not yet clear what this might mean in terms of effects, if any, on cognition and/or intelligence. Still, the brain possesses a very high energy demand, and the evolutionary trade-offs are continually adapting to new niche-specific optima aimed, ultimately, at maximizing genetic fitness utilizing the substrates available [12].

The human developmental time-line reveals that early foragers eventually became tool-makers and hunter-gatherers, and finally, as population densities increased, and there were many more mouths to feed, the first steps towards agriculture took place. The consensus among archaeologists places the advent of early agricultural practices such as the domestication of plants and animals, at approximately 10,000–12,000 years before present [13,14]. There are many theories as to the reasons some very early, but geographically distant populations, concurrently took up agriculture [15], but climate change following the last ice-age is one of the most compelling [13,16]. Generally considered a watershed moment, the adoption of agricultural practices profoundly influenced humanity, and it is typically seen as a critical step towards a better life for all. However, upon closer inspection, newly emerging techniques in paleopathology, the study of disease indicators in the remains of ancient peoples, suggest otherwise. Paleopathologists can, from ancient skeletons, calculate growth rates, determine incidences of child malnutrition, and recognize scars left on bones by anemia, tuberculosis, leprosy, and other diseases [17,18].

The transition to agriculture ultimately led to significant changes in diet after thousands of years foraging for fruit, berries, roots, wild vegetation and other edibles, and then eventually the hunting of game. It is noteworthy that the timeline related to the beginning of the agricultural revolution is centuries after the era when crucial human brain development is considered to have occurred, 2.5–2.0 mya. Hominin populations by this time consisted of larger family groups who had the cognitive capacity to communicate and cooperate with one another, to hunt, and to make rudimentary tools [19]. Indeed, *H. sapiens* have spent far longer as hunter-gatherers than as agriculturists, and the transition allowed for more permanent settlements. Still, it also led to a less diverse diet and a resultant decrease in the quality of human nutrition. As groups of hunter-gatherers switched to farming, they ultimately traded quality for quantity, and the earliest crops grown were carbohydrate-based barley, wheat, rice, and corn [17], none of which contain all of the essential amino acids or vitamins necessary for human health and survival. These crops are still farmed extensively today [20], and they make up a significant proportion of global feed and food supplies.

The increased carbohydrate content in the human diet with the shift to agriculture resulted, amongst other things, in a significant decline in dental health [17,21,22]. The relationship between dental caries and the consumption of sugar and other carbohydrates is well known, and this relationship has been used as an indicator of the dietary reconstruction as a result of agricultural intensification [23,24]. However, in 2013, Halcrow and colleagues cautioned that the carbohydrate type might have played a role, and they suggested that rice may not be particularly cariogenic. To support this standpoint, they analyzed the degree of caries in the dentition of infants and children from eight prehistoric sites in Southeast Asia. These researchers determined that while the deciduous, or baby teeth, exhibited issues related to poor dental health, the secondary, or permanent teeth, did not follow the same pattern. They concluded that while deciduous teeth were typically more susceptible to caries, the subsequent weaning of children towards a rice-based agricultural diet actually helped to maintain better oral health, contrary to evidence from regions of the world where cereals, other than rice, are utilized more extensively. An unfortunate drawback of this research relates to the unavailability of pre-agricultural samples in that region for comparison [25]. Additional declines in health potentially influenced by the adoption and transition to agriculture include a prevalence of osteoarthritis, childhood malnutrition, iron deficiency, and reduced life expectancy [17,18].

Accompanying significant changes in diet is always a corresponding change in the population diversity of gut microbes, the organisms responsible for making various enzymes and metabolites available for nutritional utilization. Alterations in macronutrient substrates available for metabolic processing create changes in nutrient supply and composition. Recent studies established that human gut microbes play multiple roles in securing the health and vitality of their host [26,27]. Mammals are metagenomic in that they possess not only their own complement of genes but also those of all of their associated microbes [28]. The contribution of the gut microbiome to the host gene pool is estimated to be over 100 times more than that of the human genome [29], and its profound influence on health and wellness is now widely recognized. The primitive human biome developed naturally in association with a variety of microbes. In an extensive sequencing study, Moeller et al., 2016 revealed that clades of the *Bacteroidaceae* and *Bifidobacteriaceae* had been maintained exclusively within host lineages across hundreds of thousands of generations, indicating robust co-speciation, and strong vertical transmission [30]. Strains of *Bifidobacteria* and *Bacteroidetes* are now known to provide significant prebiotic benefits in mammals [31,32]. Sequence analyses also provided evidence for extensive sympatry between hosts and their colonizing microbial populations [33,34]. The co-evolution of eukaryotes and their commensal, or symbiotic microbial populations played an essential role in the health and fitness of the host then, as it does now [35,36]. A significant portion of research today continues to focus on the gut microbiome, exposing the seemingly infinite number of relationships gut microbes have with their host, whether a plant, an animal, or macroalgae.

As apex consumers, humans are dependent not only on the inherent nutritional value of foods, but we are also impacted by the components that food was exposed to as it was being produced.

In addition to essential nutrition in the form of protein, carbohydrate, fat, vitamins and minerals, biologically active compounds are also necessary for optimum health, wellness, and vitality. Human health is deeply interwoven with the fabric of terrestrial agriculture, and the whole sphere of impacts is complex and far-reaching. Without agriculture, the human populations on Earth today would only exist until the food stockpiles ran out.

However, it is necessary to look upon the broader picture as a whole. In a global situation where obesity and cardiovascular and neurological diseases are at epidemic proportions and increasing, a clear assessment of the situation is warranted, even as governments begin to recognize the high costs of obesity. It is hoped that this review will provoke some thought and consideration of tools that may be naturally available in the form of macroalgae.

2. Bioactive Compounds in Macroalgae

It is well known that seaweeds naturally possess a plethora of unique and beneficial bioactive compounds [37–40], but adequate research in human clinical trials remains limited. Some information, however, has been derived from animal trials, and an example of this in terms of an agricultural food crop was demonstrated by Fan and colleagues, 2011. Studies showed increased antioxidant capacity and enhanced food quality in spinach grown with applications of a seaweed extract [41]. Conventional agricultural practices have been suggested by many researchers to contribute to the production of foods that are less nutritious than organically produced crops. However, this theory remains controversial, and the science needs to be better refined. Highlights of much of this research to date, however, demonstrated that while basic nutrition does not appear to differ significantly based upon culture technique, the production of various phytochemicals and bioactive compounds is more prevalent in organically grown crops, including those receiving seaweed-based inputs [42–45].

Research on the enhancement of antimicrobial activities of essential oils by the application of seaweed extract highlights an indirect benefit to humans. The mint and sweet basil in this study are traditional global medicinal plants grown on an industrial scale, and a foliar application of two doses weekly of *Ascophyllum nodosum* extract at 5 and 7 mL L^{-1} for 12 weeks enriched essential oil content and quality. *A. nodosum* treated plants were more productive and showed higher antibacterial properties than the control, thus providing higher quality oils without the negative environmental impacts of synthetic fertilizers [46]. For a more thorough review on the history, development, and extensive bioactive compounds found in seaweed extracts, relative to the many benefits they can afford the agricultural industry, please see Craigie, 2011 [47].

Animal-derived products such as meat, dairy, eggs, fish, and shellfish currently represent 43% of the total protein supply for human consumption [48], and this number is expected to grow in concert with global population increases. Efforts are in play to improve the nutritional quality of certain agricultural crops by biofortification methods, particularly for essential mineral elements often lacking in human diets, such as iron, zinc, copper, calcium, magnesium, and iodine [49]. Collectively, seaweeds typically contain all of these elements and are considered a dependable source of them. Reliance by societies on agricultural production means that basic nutritional requirements are being met, but what of the other important wellness compounds mentioned here previously?

The world's populations will always be dependent upon agriculture but if we consider the current situation where global health issues continue to rise, despite the availability of agriculturally produced foods, especially in industrialized countries, is something fundamental being overlooked? It is possible, perhaps even likely, that primitive hunter-gatherers enjoyed better health and wellness than people today because they foraged for a wide variety of wild foods. Coastal diets would have been optimal, enriched in phytonutrients and their associated bioactivities, including prebiotic effects. It is challenging to determine the wellness activities of ancient peoples conclusively, and in an effort to compare energy expenditure of Westernized humans to hunter-gatherer ancestors, Pontzer and colleagues, 2012 examined daily energy expenditure and physical activity levels of present-day Hadza foragers. The Hadza lifestyle is similar in critical ways to those of Pleistocene ancestors in that they hunt and gather on foot with bows, small axes, and digging sticks, and such isolated populations for study are scarce. Over 95% of the calories in the Hadza diet come from wild foods, such as tubers, berries, small and large game, baobab fruit, and honey. From this research study, the authors concluded that Hadza individuals had lower percentages of body fat than Westerners, but contrary to expectations, total energy expenditures were similar across populations. These results add to the view that energy intake is more influential than energy expenditure in relation to obesity. Highly processed, energy-dense but nutrient-poor foods are cited as the likely culprits contributing to the obesogenic effects experienced by westernized populations [50].

Furthermore, paleopathology studies provide some evidence that life expectancy at birth in the pre-agricultural community was approximately 26 years, whereas in the early post-agricultural community, it was reduced to nineteen years [17]. Wells and Stock, 2020 developed a conceptual framework based on evolutionary life history studies, and they applied it to better the understanding of how human biology changed in ancestral populations in association with the origins of agriculture. Their theory is based upon the assumption that energy in the form of food availability is finite and must be allocated in competition among the functions of maintenance, growth, reproduction, and defence. They argued that the origins of agriculture provoked trends in many components of biology, such as body size, fertility, and health status through the shifting of various trade-offs to new niche-specific optima [51]. Life history theory considers how organisms maximize their genetic fitness through harvesting resources from the environment and investing them in a suite of biological functions throughout their life-course [13].

A notable example of a high cost, high benefit trait in terms of a defense function is the inflammatory response. It is of high benefit because it can be life-saving during exposure to noxious challenges. Still, the high cost comes with the propensity of inflammatory defenses to interfere with normal functions. At the extreme, it can cause tissue damage, and even death (in severe cases of auto-immune disease). The inflammatory response is particularly sensitive to changes in relevant environmental factors such as an altered exposure to commensal and pathogenic microorganisms, changes in diet, antibiotics, stress, environmental and endogenous toxins, and physical activities. Chronic inflammation is universally associated with metabolic syndrome factors such as obesity [52], cardiovascular diseases [53], as well as neurodegenerative disorders [54], and cancer [55]. There have been many significant environmental changes in the past century, and it is proposed that the cost-benefit trade-off of the inflammatory response in modern human populations is not optimized to the current environmental situation, of which diet plays a primary role [56]. The prevalence of inflammatory diseases has increased significantly over recent decades, and the anti-inflammatory effects of seaweeds are well documented [57–60]

Efforts to improve the nutritional value of food crops typically focus on biofortification methods by the application of inorganic fertilizers, or specialized plant-breeding techniques, and the development of transgenic plants. It is also recognized that increasing the concentrations of bioactive substances in foods, such as β-carotene, cysteine-rich polypeptides, and certain organic and amino acids in foods helps to improve the bioavailability of certain nutrients [49]. Furthermore, farm animal production, including aquaculture, ruminant and monogastric livestock, is expected to increase by 70% to feed the anticipated human population increase to 9.6 billion individuals by 2050 [61]. In a comprehensive review, Garcia-Vaquero, 2019 presented an informative collection of studies which demonstrated numerous health and fitness benefits from the inclusion of seaweeds in the animal diet. Most of the studies are related to benefits to test-animals, fish, poultry, or shellfish, and are centered around improved growth as a function of the protein content of the macroalgae. Some of the beneficial responses also included pathogen resistance, enhanced immunity, and increased carotenoid content, suggesting more factors are at play beyond protein content [62].

Studies on the impact of foods on neurological health are advancing scientists' understanding of the dynamic interactions of the food-brain axis, and these studies have demonstrated that selected seaweeds contain compounds that are neuroprotective [63–66]. However, it is currently uncertain if, as secondary consumers, humans would receive the same neuroprotective benefits from agricultural crops (domesticated plants and animals), grown with the assistance of these beneficial seaweeds. Fan et al., 2011 showed that an extract from the brown seaweed, *Ascophyllum nodosum* enhanced the antioxidant content in spinach, which in turn, when prepared as a feed, protected the nematode, *Caenorhabditis elegans* against oxidative and thermal stress [41]. Furthermore, cows fed on a diet formulation containing *A. nodosum*, produced milk with increased iodine content and their gut microbial populations were altered in favour of increased beneficial microbes (*Firmicutes*), and a decrease in the number of *Proteobacteria* [67]. Laying hens with a 10% portion of the seaweed *Macrocystis pyrifera* added to their diet increased the n-3 fatty acid content of their eggs beyond that provided by sardine oil, improving lipid composition and consumer acceptance [68]. In addition, it is commonly understood that cold-water fish and shellfish that feed on algae are a reliable food source for the important polyunsaturated fatty acids, arachidonic (AA:C20:4, n-6) and docosahexaenoic acids (DHA:C22:6, n-3), critical in human health and development. Humans must obtain these fatty acids from their diet, as do the fish and shellfish, by eating the (micro or macro) algae that contain them. These are examples of nutritional benefits becoming available as they advance "up" trophic levels of the food chain from primary consumer to secondary consumer and as a result of utilizing micro/macroalgal components.

It seems evident that by improving the diversity of the foods we eat beyond what current agricultural practices offer, human health and wellness status should rebound given sufficient time. Opportunities to add robust, but currently unconventional sources such as seaweeds, to animal feeds and agricultural crops, are numerous and realistic. There is a plethora of research to support the addition of specific seaweeds, or seaweed components to domesticated livestock feeds. This would potentially lead to a healthier, nutritionally diverse food supply, which will, in many cases, ultimately carry over to the end consumer. Nutrition must reach beyond the basics of protein, fat, carbohydrate, minerals, and vitamins, to include all the other substrates that fundamentally support and facilitate an optimized state of health and wellness. There are numerous, excellent scientific reviews published on the many applications and beneficial fitness effects of administering macroalgal components to livestock feed, [69–71] as examples. There is, as well, equally, or even more robust literature highlighting nutraceutical, pharmaceutical and therapeutic benefits to humans, as a search of the internet indicates.

A wide range of studies examining the bioactive properties of seaweed supplementation in domestic livestock feed are establishing species-specific, targeted applications or effects. These constituents are referred to as 'nutricines', highlighting their benefits beyond core nutrition [72]. A sample of these will be highlighted here, beginning with a study by Wan and colleagues, (2016) who carried out a 14-week feeding trial on Atlantic salmon smolts. The red seaweed, *Palmaria palmata*, was collected in winter, washed, dried, milled, prepared and administered at three different rates, 5, 10, and 15%, by weight, and made into pellets with the remaining feed ingredients. The researchers screened for physiological changes in basic haematology, immunological indicators, hepatic markers, and whole salmon body proximate composition. Results, in general, were unremarkable, as compared to controls, with the exception of a significant decrease in alanine transaminase at the 5 and 15% inclusion rates, which is thought to be a positive indicator of liver health. In addition, at the 5% level of seaweed inclusion in the salmon feed, lipid content increased compared to control fish, and an increase in lipid concentration can reflect positively on the organoleptic characteristics of the product in addition to enhanced nutrition [73].

While the purists of the world claim that wild-caught fish are nutritionally superior to farmed fish as a competent and reliable food supply, widespread systems of aquacultured fish are essential to help feed the masses in a safe and ecologically sound manner. This will be especially true in the future, and as such an important food source, fish need to be farmed carefully and sustainably. Efforts must be made to minimize stresses and promote nutritional balance in aquaculture operations. Much attention has been placed on enhancing natural immune function in farmed fish, which also improves health and growth, reduces mortality, prevents some diseases, and increases resistance to parasites. Seaweeds are amongst the most promising immunostimulants tested to date, some shown to promote growth, stimulate appetite, enhance tonicity, improve immunity, and exhibit anti-pathogenic properties [74].

Pork is the world's most consumed meat from terrestrial animals, and global commercial pork production in 2020 was estimated, by a web-based statistics company, to be around 88.73 billion metric tons [75]. It would be wise to ensure that if such a significant amount of pork is being consumed, then that meat should not only be a source of protein, fat, and specific vitamins, but should also contribute important phytocompounds. This is a potentially beneficial way to contribute to global health. However, the research is far from complete, although reported here are some examples utilizing brown seaweeds or brown seaweed extracts for livestock.

The polysaccharides laminarin and fucoidan are present in brown seaweeds, and these compounds provide important anti-inflammatory and prebiotic effects in livestock. Post-weaning pigs were fed for two weeks a diet supplemented with a 300 mg/kg laminarin-rich extract sourced from BioAtlantis

Ltd. (Clash Industrial Estate, Tralee, Co. Kerry, Ireland). The pigs had higher feed intake, growth rate, and body weight, as compared to controls. Laminarin is found in kelp-like seaweeds and is especially abundant in *Saccharina/Laminaria* spp. Corresponding to the improved fitness of the treated pigs was a measured proliferation of bacterial taxa. These included *Prevotella* that favourably enhanced nutrient digestion whilst reducing the load of potentially pathogenic bacterial taxa, including *Enterobacteriaceae* [76]. Laminarin is an important bioactive compound, and it is a robust source of antioxidants as well as an algal source of β-glucan, a natural compound known for its purported functionality in foods and its immunity-enhancing properties. However, the exact mechanisms are not yet fully understood [77]. When researchers supplemented weanling pig diets with 2.5% powdered *Laminaria* as a component of the basal feed formulation, they found that the seaweed additive not only served a nutritional purpose, but that it also exerted additional bioactivities with a positive impact on productivity [78].

Dried and ground stalks (stipes), of the brown seaweed *Undaria pinnatifida,* when fed to pigs at a rate of 1%, altered their intestinal microflora preferentially in favour of probiotic populations, such as *Lactobacillus* spp. for example, and were inhibitory to pathogens such as *Escherichia coli*. In addition, immunomodulatory effects, demonstrated by a significantly higher percentage of peripheral, blood natural-killer-cells in the treatment groups, is a promising step towards the reduction of widespread antibiotic usage [79].

Brown seaweeds, in general, are an excellent source of bioactive compounds. However, a point to bear in mind is that there is an inherent variability within and amongst species, and samples must be well characterized. In addition, there is a temptation to consider that if a little bit of something is good, then a lot should be better, and care must be taken to ensure the science is complete and thorough. Some studies showed that when certain seaweeds were used as a feed replacer, and therefore administered in higher doses, negative results such as scours and loss of conditioning in farmed animals occurred [72]. This possibility has led researchers to refine their approach and consider using seaweeds for their potent prebiotic effects as lower dose feed supplements also aimed at the potential synergies associated with whole, usually granulated seaweeds. For example, in a detailed review highlighting the benefits of macroalgae in poultry feed, it was reported that the addition of 0.5 kg of *A. nodosum* per metric ton of feed significantly reduced the effects of prolonged heat stress on the birds [80]. Certain seaweed extracts such as fucoidan and laminarin, which can focus on specific activities, are also of interest for targeted applications such as enhanced immunity and improved gut health [32,81]. Still, in consideration of poultry, Kulshreshtha and colleagues, investigated in 2020 seaweed components as agents against the drug-resistant pathogen *Salmonella Enteritidis*, carrying out the study in cell cultures. They investigated the effects of water extracts of two cultivated red seaweeds, *Chondrus crispus*, and *Sarcodiotheca gaudichaudii* in various combinations with industry-standard antibiotics. Streptomycin exhibited higher antimicrobial activity against S. Enteritidis compared to tetracycline, with a MIC_{25} and MIC_{50} of 1.00 and 1.63 µg/mL, respectively. However, the addition of a water extract of *C. crispus* at a concentration of 200 µg/mL to the tetracycline treatment significantly enhanced antibacterial activity (log CFU/mL 4.7 and 4.5 at MIC_{25} and MIC_{50}, respectively). Furthermore, the *S. gaudichaudii* water extract, at 400 and 800 mg/mL, and also in combinations with tetracycline, showed total inhibition of bacterial growth [82]. The reduction of antibiotics in global agricultural operations is a very important step towards minimizing resistance to synthetic drugs, and overall, by extension, will help lead to the development of a healthier planet for people, plants, and animals naturally.

Researchers continue to explore the benefits of supplementing livestock diets with seaweeds, and whilst protein substitution is one of the more popular reasons for developing alternative feedstuffs, other vital applications are coming to light with respect to the meat and dairy industries. A sun-dried, specially managed granular extract of the fucoid, *A. nodosum* (Tasco™, Acadian Seaplants Ltd., Dartmouth, NS, Canada) was fed at a 1, 3, or 5% rate to young rams for 21 days. There was no effect under the conditions of this study on rumen fermentation, but rumen total bacteria and archaea were linearly reduced, and protozoa were linearly increased by increasing levels of Tasco™. Furthermore, the addition of seaweed to feed reduced the total *E. coli* population, a common and ubiquitous, foodborne pathogen [83]. Other examples of the positive impact seaweeds can induce include benefits to cattle, and steers fed 20 g Tasco™/kg diet for seven days showed similar effects in pathogen reduction. Fecal shedding of *E. coli* O157:H7 was significantly reduced in both duration and intensity, indicating an inhibitory effect on the growth and proliferation of this virulent bacteria [84]. This is a valid example of a source of contamination coming from animals and negatively impacting humans, as frequent meat recalls for *E. coli* can attest.

Addition of powdered *A. nodosum* (80.0 g/cow) to feed for Holstein dairy cows increased blood glucose levels, and it decreased blood sorbitol dehydrogenase (SDH) levels. The activity of SDH in the blood of healthy animals is low, whereas its elevation above normal range implies hepato-cellular injury. The authors suggested that this result may indicate a hepatoprotective effect of the seaweed, in concert with improved energy utilization [85].

An unfortunate reality of cattle production is evidence that in the process of digesting and utilizing their feed components, bacterial fermentation in their gut releases significant amounts of methane, a particularly undesirable greenhouse gas. Of all the reported livestock produced in the world, cattle contribute approximately 62% of global emissions within the animal sector [86], and efforts to find ways to reduce this inherent methane production have become a priority. Dietary seaweeds have been shown repeatedly to influence the gut microbiome, and this holds true as well for the methane-producing microbes, including members of the Archaea. However, there remains extensive variability in effectiveness, based upon seaweed species and inclusion rates [87]. Although the rumen microbiome can ferment seaweeds and provide energy to the host animal, high variability of digestibility values is evident among and within seaweed species, and this applies to the methane-reduction effects as well [88]. Much more research is currently required, but one of the most promising seaweeds to date, for the promotion of anti-methanogenesis activity in cattle, is the red algae *Asparagopsis taxiformis*, which provided over 90% methane reduction, at a supplement level of 2% organic matter in in vitro trials [89,90]. Developing safe and effective methods for the reduction of enteric methanogenesis is indeed, challenging, and any adoptive strategies need to be sustainable, practical, and economically viable. Whilst methane reduction is not a direct nutritional benefit for humans, it does have far-reaching and important global climatic implications, and further studies on a wide range of seaweed-based dietary supplements should be undertaken.

The red seaweed *Chondrus crispus* has a long history of usage as food and medicine [91], and this species is remarkable in its bioactive characteristics, as demonstrated by various animal studies. Components of this seaweed were shown to enhance host immunity, suppressing the expression of quorum sensing and the virulence factors of a *Pseudomonas aeruginosa* strain, and enrich probiotic levels in the host [92,93]. As a supplemental feed ingredient, 2% *C. crispus* significantly increased the beneficial (probiotic) bacteria in the guts of layer hens, and it also enriched the short-chain fatty acid concentration, which is thought to act as an energy source for intestinal epithelial cells, stimulating cell growth [94]. Sulphated polysaccharides (SP) were extracted from samples of *C. crispus* collected off the Atlantic coast of Ireland and used to determine the effects, if any, on wild mussels. Results indicated that the SP from *C. crispus* rapidly induced health-enhancing activities in *Mytilus* spp. at a cellular, humoral and molecular level, and with up to a 10-day prolonged effect [95].

While scientific research regarding the impact of seaweed on equine health is somewhat limited, there is an important accumulation of anecdotal evidence for the utilization of microelements, conditioning, and other benefits derived from seaweeds. Interestingly, excessive obesity affects approximately 45% of the worldwide horse population, resulting in equine metabolic syndrome (EMS) [96], which parallels metabolic syndrome (MetS) in the human population. The same fundamentals of nutrition apply to all, and equine also utilize basic substrates to optimize genetic fitness. Still, the benefits for horses of nutricines from seaweeds have yet to be thoroughly investigated. There are several macroalgal-based products currently in the marketplace, primarily as supplements, and quality testing must be done for these to meet the appropriate regulatory criteria [97]. Horses are not a widely used food animal for humans, but they constitute a significant proportion of agricultural livestock, and they make a large contribution to feed consumption statistics. Their general health status is but another example of the impact domestication has had on the world's livestock.

3. Conclusions—One IS what One Eats

It is obviously unreasonable to consider that all the members of the global, human population may return to a lifestyle of foraging, hunting, and fishing, for the acquisition of a nutritionally balanced, wild-based diet. From archeological studies, including paleopathology, it is now understood that many of the *H. sapiens* on the planet, prior to the advent of agriculture, in fact, had a more diverse diet, and are considered to have had healthier lives. In addition, as mentioned previously, the ancestors of today's humans must have had long-term access to all the essential nutrients for the growth and development of the brain with respect to its structure and sophistication—characteristics that differentiate humans from other primates [6]. Simply as a function of accessing natural food sources, useful bioactive compounds, such as antioxidants, various polyphenols, and also in the case of seaweeds, sulfated polysaccharides would be present and plentiful.

In a well-cited commentary (~2500 citations on 20 June, 2020) published in 2005, the authors remarked, "the evolutionary collision of our ancient genome with the nutritional qualities of recently introduced foods may underlie many of the chronic diseases of Western civilizations", and they provide several reasons for that statement. In particular, lower nutritional diversity, extensive food processing, and the over-consumption of sugar and salt are cited as primary factors negatively impacting the health status of global populations [98]. Without a doubt, the industrialization of the food system has further reduced dietary variability. As an example, the genetic diversity of the corn plant has now been lost, but the number of products made from this crop is in the thousands [99]. Would growing that corn with seaweed extract lead to a broader range of antioxidants for those who consume it? How many steps in the food web would maintain that bioactive nutrition? What measures must be taken to improve global health? Big questions, with complicated answers, even without addressing the various economic scenarios and their influence. It is a fact, however, that chronic diseases are impacting populations across the world, and they continue to rise, disrupting such vital organs as hearts and brains, and contributing to diabetes, obesity, inflammatory disorders, and loss of vitality. Furthermore, many of these conditions are interconnected, where one negatively influences the other, and now studies reveal that there is a strong relationship with the gut microbiome, which has co-evolved along with humans, other animals, and plants. Host dietary intake is a major environmental factor influencing gut bacterial abundances and disease phenotypes [100], and obviously, the phytonutrient quality of our food matters—especially over the longer term.

The solution, therefore, must lie within our current systems of food production, and the quality and diversity of the human diet must be improved upon. It seems quite possible that by adding seaweeds, or seaweed supplements to animal feeds, that at least some beneficial compounds will carry over to the end consumer. However, just as there are a plethora of bioactive metabolites in seaweeds, there is also a plethora of variables that must be identified and managed, and this will require significantly more research. Species differences, dosages, economics and raw material supplies must all be considered. Bioavailability of active compounds may also be a challenge as scientists strive to understand how functional constituents interact metabolically within their host [101], and potentially there will also be some individual differences in terms of response. As an aquatic natural resource, macroalgae are influenced by the quality of the environment in which they grow, and therefore it is important to be cognizant of any potential toxins or antinutrients, such as excess heavy metals.

There is no escaping the fact that with respect to food, one is, what one eats. Efforts to enrich our agricultural products by supplementing with seaweed, which contains a wide variety of phytonutrients (perhaps, 'phyconutrients' in this case) are a realistic and practical way to begin diversifying diets. There is some concern that the rates of whole dried seaweed applied may not be enough to cause a desired effect, whereas too much may have palatability or other issues. Step-wise approaches to significant or long-term dietary changes may enable rumen microbes in cattle, for example, to adapt to higher doses over time if necessary, to obtain optimized nutritional value [70]. As an ancient food source, available thousands of years before terrestrial plants, it would seem prudent to increase the agricultural use of seaweeds as feed and food, with the aim of improving global human health. Obviously, there is some distance to go to reach that point. Still, competent and dedicated researchers are working on this opportunity, and with the eventual implementation of seaweed-derived foods and feeds into agriculture, the next 10,000 years will result in a much healthier planet, including the beings living here.

Author Contributions: M.L.C. and A.T.C.; formal analysis, M.L.C.; investigation, M.L.C.; writing—original draft preparation, M.L.C. and A.T.C.; writing—review and editing, M.M.M.; visualization, M.L.C. and A.T.C.; supervision, A.T.C. All authors have read and agreed to the published version of the manuscript.

Funding: This research received no external funding.

Acknowledgments: The authors are most grateful to the anonymous reviewers for their valuable and constructive comments, and also to those dedicated phycologists whom so diligently research seaweeds, with a conscientious goal to positively impact detrimental global issues.

Conflicts of Interest: The authors declare no conflict of interest.

References

1. Sistiaga, A.; Wrangham, R.; Rothman, J.M.; Summons, R.E. New insights into the evolution of the human diet from faecal biomarker analysis in wild chimpanzee and gorilla faeces. *PLoS ONE* **2015**, *10*, e0128931. [CrossRef] [PubMed]

2. Cerling, T.E.; Manthi, F.K.; Mbua, E.N.; Leakey, L.N.; Leakey, M.G.; Leakey, R.E.; Brown, F.H.; Grine, F.E.; Hart, J.A.; Kaleme, P.; et al. Stable isotope-based diet reconstructions of Turkana Basin hominins. *Proc. Natl. Acad. Sci. USA* **2013**, *110*, 10501–10506. [CrossRef] [PubMed]

3. Sponheimer, M.; Alemseged, Z.; Cerling, T.E.; Grine, F.E.; Kimbel, W.H.; Leakey, M.G.; Lee-Thorp, J.A.; Manthi, F.K.; Reed, K.E.; Wood, B.A.; et al. Isotopic evidence of early hominin diets. *Proc. Natl. Acad. Sci. USA* **2013**, *110*, 10513–10518. [CrossRef]

4. Maberly, S.C.; Raven, J.A.; Johnston, A.M. Discrimination between ^{12}C and ^{13}C by marine plants. *Oecologia* **1992**, *91*, 481–492. [CrossRef]

5. Bumstead, M.P. The potential of stable carbon isotopes in bioarcheological anthropology. Research Report 20: Biocultural Adaptation Comprehensive Approaches to Skeletal Analysis, 1981, p 12. Available online: https://core.ac.uk/display/13640207 (accessed on 19 May 2020).

6. Cornish, M.L.; Mouritsen, O.G.; Critchley, A.T. Consumption of seaweeds and the human brain. *J. Appl. Phycol.* **2017**, *29*, 2377–2398. [CrossRef]

7. Aiello, L.C.; Wheeler, P. The expensive tissue hypothesis: The brain and the digestive system in human and primate evolution. *Curr. Anthropol.* **1995**, *36*, 199–221. [CrossRef]

8. Hawks, J.; Hunley, K.; Lee, S.-H.; Wolpoff, M. Population bottlenecks and Pleistocene human evolution. *Mol. Biol. Evol.* **2000**, *17*, 2–22. [CrossRef]

9. Glazko, G.V.; Nei, M. Estimation of divergence times for major lineages of primate species. *Mol. Biol. Evol.* **2003**, *20*, 424–434. [CrossRef]

10. Henneberg, M. Decrease of human skull size in the Holocene. *Human Biol.* **1988**, *60*, 395–405.

11. Hawks, J. Selection for smaller brains in Holocene human evolution. *arXiv* **2011**, arXiv:abs/1102.5604.

12. Hill, K. Life history theory and evolutionary anthropology. *Evol. Anthropol.* **1993**, *2*, 78–89. [CrossRef]

13. Gupta, A.K. Origin of agriculture and domestication of plants and animals linked to early Holocene climate amelioration. *Curr. Sci.* **2004**, *87*, 54–59.

14. Riehl, M.; Zeidi, M.; Conard, N.J. Emergence of agriculture in the foothills of the Zagros Mountains of Iran. *Science* **2013**, *341*, 65–67. [CrossRef] [PubMed]

15. Neolithic Revolution. Available online: https://en.wikipedia.org/wiki/Neolithic_Revolution (accessed on 30 July 2020).

16. Broushaki, F.; Thomas, M.G.; Link, V.; López, S.; van Dorp, L.; Kirsanow, K.; Hofmanová, Z.; Diekmann, Y.; Cassidy, L.M.; Díez-del-Molino, D.; et al. Early Neolithic genomes from the eastern Fertile Crescent. *Science* **2016**, *353*, 499–503. [CrossRef] [PubMed]

17. Cohen, M.N.; Armelagos, G.J. (Eds.) *Paleopathology at the Origins of Agriculture*; Academic Press: New York, NY, USA, 1984.

18. Larsen, C.S. Biological changes in human populations with agriculture. *Annu. Rev. Anthropol.* **1995**, *24*, 185–213. [CrossRef]

19. Elton, S. The environmental context of human evolutionary history in Eurasia and Africa. *J. Anat.* **2008**, *212*, 377–393. [CrossRef]

20. United States Department of Agriculture (USDA). World Agriculture Production. 2020. Available online: https://www.fas.usda.gov/data/world-agricultural-production (accessed on 20 June 2020).

21. Turner, C.G., II. Dental anthropological indications of agriculture among the Jomon people of central Japan. X. Peopling of the Pacific. *Am. J. Phys. Anthropol.* **1979**, *51*, 619–636. [CrossRef]

22. Temple, D.H.; Larsen, C.S. Dental caries prevalence as evidence for agriculture and subsistence variation during the Yayoi period in prehistoric Japan: Biocultural interpretations of an economy in transition. *Am. J. Phys. Anthropol.* **2007**, *134*, 501–512. [CrossRef]

23. Molnar, S.; Molnar, I. Observations of dental diseases among prehistoric populations of Hungary. *Am. J. Phys. Anthropol.* **1985**, *67*, 51–63. [CrossRef]

24. Larsen, C.S. *Bioarcheology: Interpreting Behaviour from the Human Skeleton*, 2nd ed.; Cambridge University Press: Cambridge, UK, 2015; p. 654.

25. Halcrow, S.E.; Harris, N.J.; Tayles, N.; Ikehara-Quebral, R.; Pietrusewsky, M. From the mouths of babes: Dental caries in infants and children and the intensification of agriculture in mainland Southeast Asia. *Am. J. Phys. Anthropol.* **2013**, *150*, 409–420. [CrossRef]

26. Donia, M.S.; Cimermancic, P.; Schulze, C.J.; Brown, L.C.W.; Martin, J.; Mitreva, M.; Clardy, J.; Linington, R.G.; Fischbach, M.A. A systematic analysis of biosynthetic gene clusters in the human microbiome reveals a common family of antibiotics. *Cell* **2014**, *158*, 1402–1414. [CrossRef] [PubMed]

27. Cani, P.D.; Everard, A. Talking microbes: When gut bacteria interact with diet and host organs. *Mol. Nutr. Food Res.* **2016**, *60*, 58–66. [CrossRef] [PubMed]

28. Ley, R.E.; Hamady, M.; Lozupone, C.; Turnbaugh, P.J.; Ramey, R.R.; Bircher, J.S.; Schlegel, M.L.; Tucker, T.A.; Schrenzel, M.D.; Knight, R.; et al. Evolution of mammals and their gut microbes. *Science* **2008**, *320*, 1647–1651. [CrossRef] [PubMed]

29. Qin, J.J.; Li, R.Q.; Raes, J.; Arumugam, M.; Burgdorf, K.S.; Manichanh, C.; Nielsen, T.; Pons, N.; Levenez, F.; Yamada, T.; et al. A human gut microbial gene catalogue established by metagenomic sequencing. *Nature* **2010**, *464*, 59–65. [CrossRef] [PubMed]

30. Moeller, A.H.; Caro-Quintero, A.; Mjungu, D.; Georgiev, A.V.; Lonsdorf, E.V.; Muller, M.N.; Pusey, A.E.; Peeters, M.; Hahn, B.H.; Ochman, H. Cospeciation of gut microbiota with hominids. *Science* **2016**, *353*, 380–382. [CrossRef]

31. Mariat, D.; Firmesse, O.; Levenez, F.; Guimaraes, V.D.; Sokol, H.; Dore, J.; Corthier, G.; Furet, J. The Firmicutes/Bacteroidetes ratio of the human microbiota changes with age. *BMC Microbiol.* **2009**, *9*, 123. [CrossRef]

32. You, L.; Gong, Y.; Li, L.; Hu, X.; Brennan, C.; Kulikouskaya, V. Beneficial effects of three brown seaweed polysaccharides on gut microbiota and their structural characteristics: An overview. *Inter. J. Food Sci. Technol.* **2019**, *55*, 1199–1206. [CrossRef]

33. Sonnenburg, J.L. Genetic potluck. *Nature* **2010**, *464*, 837–838. [CrossRef]

34. Blaser, M.J. The past and future biology of the human microbiome in an age of extinctions. *Cell* **2018**, *172*, 1173–1177. [CrossRef]

35. Margulis, L. *Symbiosis in Cell Evolution: Microbial Communities in the Achean and Proterozoic Eons*, 2nd ed.; Freeman: New York, NY, USA, 1993.

36. Zilber-Rosenburg, I.; Rosenburg, E. Role of microorganisms in the evolution of animals and plants: The hologenome theory of evolution. *FEMS Microbiol. Rev.* **2008**, *32*, 723–735. [CrossRef]

37. Freile-Pelegrín, Y.; Robledo, D. Bioactive phenolic compounds from algae. In *Bioactive Compounds from Marine Foods*; Hernández-Ledesma, B., Herrero, M., Eds.; John Wiley and Sons Ltd.: New York, NY, USA, 2013; pp. 113–129. [CrossRef]

38. Pal, A.; Kamthania, M.C.; Kumar, A. Bioactive compounds and properties of seaweeds—A review. *Open Access Libr. J.* **2014**, *1*, e752. [CrossRef]

39. Holdt, S.L.; Kraan, S. Bioactive compounds in seaweed: Functional food application and legislation. *J. Appl. Phycol.* **2011**, *23*, 543–597. [CrossRef]

40. Schepers, M.; Martens, N.; Tiane, A.; Vanbrabant, K.; Liu, H.-B.; Lütjohann, D.; Mulder, M.; Vanmierlo, T. Edible seaweed-derived constituents: An undisclosed source of neuroprotective compounds. *Neur. Regen. Res.* **2020**, *15*, 790–795.

41. Fan, D.; Hodges, D.M.; Zhang, J.; Kirby, C.W.; Ji, X.; Locke, S.J.; Critchley, A.T.; Prithiviraj, B. Commercial extract of the brown seaweed *Ascophyllum nodosum* enhances phenolic content of spinach (*Spinacia oleracea* L.) which protects *Caenorhabditis elegans* against oxidative and thermal stress. *Food Chem.* **2011**, *124*, 195–202. [CrossRef]

42. Brandt, K.; Mølgaard, J.S. Organic agriculture: Does it enhance or reduce the nutritional value of plant foods? *J. Sci. Food Agric.* **2001**, *81*, 924–931. [CrossRef]

43. Wang, S.Y.; Chen, C.-T.; Sciarappa, W.; Wang, C.Y.; Camp, M.J. Fruit quality, antioxidant capacity, and flavonoid content of organically and conventionally grown blueberries. *J. Agric. Food Chem.* **2008**, *56*, 5788–5794. [CrossRef] [PubMed]

44. Hallmann, E. The influence of organic and conventional cultivation systems on the nutritional value and content of bioactive compounds in selected tomato types. *J. Sci. Food Agric.* **2012**, *92*, 2840–2848. [CrossRef] [PubMed]

45. Vinha, A.F.; Barreira, S.V.P.; Costa, A.S.G.; Alves, R.C.; Oliveira, M.B.P.P. Organic versus conventional tomatoes: Influence on physicochemical parameters, bioactive compounds and sensorial attributes. *Food Chem. Toxicol.* **2014**, *67*, 139–144. [CrossRef] [PubMed]

46. Elansary, H.O.; Yessoufou, K.; Shokralla, S.; Mahmoud, E.A.; Skalicka-Woźniak, K. Enhancing mint and basil oil composition and antibacterial activity using seaweed extracts. *Indust. Crops Prod.* **2016**, *92*, 50–56. [CrossRef]

47. Craigie, J.S. Seaweed extract stimuli in plant science and agriculture. *J. Appl. Phycol.* **2011**, *23*, 371–393. [CrossRef]

48. Henchion, M.; Hayes, M.; Mullen, A.M.; Fenelon, M.; Tiwari, B. Future protein supply and demand: Strategies and factors influencing a sustainable equilibrium. *Foods* **2017**, *6*, 53. [CrossRef] [PubMed]

49. White, P.J.; Broadley, M.R. Biofortification of crops with seven mineral elements often lacking in human diets—Iron, zinc, copper, calcium, magnesium, selenium and iodine. *New Phytol.* **2009**, *182*, 49–84. [CrossRef] [PubMed]

50. Pontzer, H.; Raichlen, D.A.; Wood, B.M.; Mabulla, A.Z.P.; Racette, S.B.; Marlowe, F.W. Hunter-gatherer energetics and human obesity. *PLoS ONE* **2012**, *7*, e40503. [CrossRef] [PubMed]

51. Wells, J.C.K.; Stock, J.T. Life history transitions at the origins of agriculture: A model for understanding how niche construction impacts human growth, demography and health. *Fron. Endocrin.* **2020**, *11*, 325. [CrossRef]

52. Hotamisligil, G.S.; Erbay, E. Nutrient sensing and inflammation in metabolic diseases. *Nat. Rev. Immunol.* **2008**, *8*, 923–934. [CrossRef]

53. Libby, P. Inflammation and cardiovascular disease mechanisms. *Am. J. Clin. Nut.* **2006**, *83*, 456S–460S. [CrossRef]

54. Wyss-Coray, T.; Mucke, L. Inflammation in neurodegenerative disease—A double-edged sword. *Neuron* **2002**, *35*, 419–432. [CrossRef]

55. Trinchieri, G. Cancer and inflammation: An old intuition with rapidly evolving new concepts. *Annu. Rev. Immunol.* **2012**, *30*, 677–706. [CrossRef]

56. Okin, D.; Medzhitov, R. Evolution of inflammatory diseases. *Curr. Biol.* **2012**, *22*, R733–R740. [CrossRef]

57. Dutot, M.; Fagon, R.; Hemon, M.; Rat, P. Antioxidant, anti-inflammatory, and anti-senescence activities of a phlorotannin-rich natural extract from brown seaweed *Ascophyllum nodosum*. *Appl. Biochem. Biotechnol.* **2012**, *167*, 2234–2240. [CrossRef]

58. Robertson, R.C.; Guihéneuf, F.; Bahar, B.; Schmid, M.; Stengel, D.B.; Fitzgerald, G.F.; Ross, R.P.; Stanton, C. The anti-inflammatory effect of algae-derived lipid extracts on lipopolysaccharide (LPS)-stimulated human THP-1 macrophages. *Mar. Drugs* **2015**, *13*, 5402–5424. [CrossRef] [PubMed]

59. Bahar, B.; O'Doherty, J.V.; Smyth, T.J.; Sweeney, T. A comparison of the effects of an *Ascophyllum nodosum* ethanol extract and its molecular weight fractions on the inflammatory immune gene expression In-Vitro and Ex-Vivo. *Innov. Food Sci. Emer. Technol.* **2016**, *37*, 276–285. [CrossRef]
60. Yokota, T.; Nomura, K.; Nagashima, M.; Kamimura, N. Fucoidan alleviates high-fat diet-induced dyslipidemia and atherosclerosis in ApoE ^{shl} mice deficient in apolipoprotein E expression. *J. Nutr. Biochem.* **2016**, *32*, 46–54. [CrossRef] [PubMed]
61. Herman, E.M.; Schmidt, M.A. The potential for engineering enhanced functional-feed soybeans for sustainable aquaculture feed. *Front. Plant Sci.* **2016**, *7*, 440. [CrossRef]
62. Garcia-Vaquero, M. Seaweed proteins and applications in animal feed. In *Novel Proteins for Food, Pharmaceuticals, and Agriculture: Sources, Applications, and Advances*; Hayes, M., Ed.; John Wiley and Sons: Hoboken, NJ, USA, 2019; pp. 139–161.
63. Gao, Y.; Dong, C.; Yin, J.; Shen, J.; Tian, J.; Li, C. Neuroprotective effect of fucoidan on H_2O_2-induced apoptosis in PC12 cells via activation of PI3K/Akt pathway. *Cell Mol. Neurobiol.* **2012**, *32*, 523–529. [CrossRef]
64. Barbosa, M.; Valentão, P.; Andrade, P.B. Bioactive compounds from macroalgae in the new millennium: Implications for neurodegenerative diseases. *Mar. Drugs* **2014**, *12*, 4934–4972. [CrossRef]
65. Sangha, J.S.; Wally, O.; Banskota, A.H.; Stefanova, R.; Hafting, J.T.; Critchley, A.T.; Prithiviraj, B. A cultivated form of a red seaweed (*Chondrus crispus*) suppresses β-amyloid-induced paralysis in *Caenorhabditis elegans*. *Mar. Drugs* **2015**, *13*, 6407–6424. [CrossRef]
66. Olasehinde, T.A.; Olaniran, A.O.; Okoh, A.I. Macroalgae as a valuable source of naturally occurring bioactive compounds for the treatment of Alzheimer's Disease. *Mar. Drugs* **2019**, *17*, 609. [CrossRef]
67. Lopez, C.C.; Serio, A.; Rossi, C.; Mazzarrino, G.; Marchetti, S.; Castellani, F.; Grotta, L.; Fiorentino, F.P.; Paparella, A.; Martino, G. Effect of diet supplemented with *Ascophyllum nodosum* on cow milk composition and microbiota. *J. Dairy Sci.* **2016**, *99*, 6285–6297. [CrossRef]
68. Carrillo, S.; López, E.; Casas, M.M.; Avila, E.; Castillo, R.M.; Carranco, M.E.; Calvo, C.; Pérez-Gil, F. Potential use of seaweeds in the laying hen ration to improve the quality of n-3 fatty acid enriched eggs. *J. Appl. Phycol.* **2008**, *20*, 721–728. [CrossRef]
69. Rajauria, G.; Cornish, L.; Ometto, F.; Msuya, F.E.; Villa, R. Identification and selection of algae for food, feed, and fuel applications. In *Seaweed Sustainability*; Tiwari, B.K., Troy, D.J., Eds.; Academic Press: London, UK, 2015; pp. 315–345.
70. Makkar, H.P.S.; Tran, G.; Heuzé, V.; Giger-Reverdin, S.; Lessire, M.; Lebas, F.; Ankers, P. Seaweeds for livestock diets: A review. *Anim. Feed Sci. Technol.* **2016**, *212*, 1–17. [CrossRef]
71. Øverland, M.; Mydland, L.T.; Skrede, A. Marine macroalgae as sources of protein and bioactive compounds in feed for monogastric animals. *J. Sci. Food Agric.* **2019**, *99*, 13–24. [CrossRef] [PubMed]
72. Evans, F.D.; Critchley, A.T. Seaweeds for animal production use. *J. Appl. Phycol.* **2014**, *26*, 891–899. [CrossRef]
73. Wan, A.H.L.; Soler-Vila, A.; O'Keeffe, D.; Casburn, P.; Fitzgerald, R.; Johnson, M.P. The inclusion of *Palmaria palmata* macroalgae in Atlantic salmon (*Salmo salar*) diets: Effects on growth, haematology, immunity, and liver function. *J. Appl. Phycol.* **2016**, *28*, 3091–3100. [CrossRef]
74. Afonso, C.P.C.N.; Mouga, T.M.L.d.S. Seaweeds as fish feed additives. In *Seaweeds as Plant Fertilizer, Agricultural Biostimulants and Animal Fodder*; Pereira, L., Bahcevandziev, K., Joshi, N.H., Eds.; CRC Press: Boca Raton, FL, USA, 2019; pp. 150–186.
75. Global Pork Production in 2020, by Country. Available online: https://www.statista.com/statistics/273232/net-pork-production-worldwide-by-country/ (accessed on 30 July 2020).
76. Vigors, S.; O'Doherty, J.V.; Rattigan, R.; McDonnell, M.J.; Rajauria, G.; Sweeney, T. Effect of a laminarin rich macroalgal extract on the caecal and colonic microbiota in the post-weaned pig. *Mar. Drugs* **2020**, *18*, 157. [CrossRef] [PubMed]
77. Ahmad, A.; Anjum, F.M.; Zahoor, T.; Nawaz, H.; Dilshad, S.M.R. Beta Glucan: A valuable functional ingredient in foods. *Crit. Rev. Food Sci. Nutr.* **2012**, *52*, 201–212. [CrossRef]
78. Brugger, D.; Bolduan, C.; Becker, C.; Buffler, M.; Zhao, J.; Windisch, W.M. Effects of whole plant brown algae (*Laminaria japonica*) on zootechnical performance, apparent total tract digestibility, faecal characteristics and blood plasma urea in weaned piglets. *Arch. Anim. Nutr.* **2020**, *14*, 19–38. [CrossRef]
79. Shimazu, T.; Borjigin, L.; Katoh, K.; Roh, S.-G.; Kitazawa, H.; Abe, K.; Suda, Y.; Saito, H.; Kunii, H.; Nihei, K.; et al. Addition of Wakame seaweed (*Undaria pinnatifida*) stalk to animal feed enhances immune response and improves intestinal microflora in pigs. *Anim. Sci. J.* **2019**, *90*, 1248–1260. [CrossRef]

80. Kulshreshtha, G.; Hincke, M.T.; Prithiviraj, B.; Critchley, A. A review of the varied uses of macroalgae as dietary supplements in selected poultry with special reference to laying hens and broiler chickens. *J. Mar. Sci. Eng.* **2020**, *8*, 536. [CrossRef]

81. Heim, G.; Sweeney, T.; O'Shea, C.J.; Doyle, D.N.; O'Doherty, J.V. Effect of maternal dietary supplementation of laminarin and fucoidan, independently or in combination, on pig growth performance and aspects of intestinal health. *Anim. Feed Sci. Technol.* **2015**, *204*, 28–41. [CrossRef]

82. Kulshreshtha, G.; Critchley, A.; Rathgeber, B.; Stratton, G.; Banskota, A.H.; Hafting, J.; Prithiviraj, B. Antimicrobial effects of selected, cultivated red seaweeds and their components in combination with tetracycline, against poultry pathogen *Salmonella* Enteritidis. *J. Mar. Sci. Eng.* **2020**, *8*, 511. [CrossRef]

83. Zhou, M.; Hünerberg, M.; Chen, Y.; Reuter, T.; McAllister, T.A.; Evans, F.; Critchley, A.T.; Guan, L.L. Air-dried brown seaweed, *Ascophyllum nodosum*, alters the rumen microbiome in a manner that changes rumen fermentation profiles and lowers the prevalence of foodborne pathogens. *mSphere* **2018**, *3*, e00017–e00018. [CrossRef] [PubMed]

84. Bach, S.J.; Wang, Y.; McAllister, T.A. Effect of feeding sundried seaweed (*Ascophyllum nodosum*) on fecal shedding of *Escherichia coli* O157:H7 by feedlot cattle and on growth performance of lambs. *Anim. Feed Technol.* **2008**, *142*, 17–32. [CrossRef]

85. Karatzia, M.; Christaki, E.; Bonos, E.; Karatzias, C.; Florou-Paneri, P. The influence of dietary Ascophyllum nodosum on haematologic parameters of dairy cows. *Ital. J. Anim. Sci.* **2016**, *11*. [CrossRef]

86. Food and Agriculture Organization (FAO). Global Livestock Environmental Assessment Model. 2020. Available online: http://www.fao.org/gleam/results/en/#c300947 (accessed on 15 June 2020).

87. Molina-Alcaide, E.; Carro, M.D.; Roleda, M.Y.; Weisbjerg, M.R.; Lind, V.; Novoa-Garrido, M. In Vitro ruminal fermentation and methane production of different seaweed species. *Anim. Feed Sci. Technol.* **2017**, *228*, 1–12. [CrossRef]

88. Cabrita, A.R.J.; Valente, I.M.; Oliveira, H.M.; Fonseca, A.J.M.; Maia, M.R.G. Effects of feeding with seaweeds on ruminal fermentation and methane production. In *Seaweeds as Plant Fertilizer, Agricultural Biostimulants and Animal Fodder*; Pereira, L., Bahcevandziev, K., Joshi, N.H., Eds.; CRC Press: Boca Raton, FL, USA, 2019; pp. 187–209.

89. Machado, L.; Magnusson, M.; Paul, N.A.; Kinley, R.; de Nys, R.; Tomkins, N. Identification of bioactives from the red seaweed *Asparagopsis taxiformis* that promote antimethanogenic activity In Vitro. *J. Appl. Phycol.* **2016**, *28*, 3117–3126. [CrossRef]

90. Machado, L.; Tomkins, N.; Magnusson, M.; Midgley, D.J.; de Nys, R.; Rosewarne, C.P. In Vitro response of rumen microbiota to the antimethanogenic red macroalga *Asparagopsis taxiformis*. *Microb. Ecol.* **2018**, *75*, 811–818. [CrossRef]

91. Craigie, J.S.; Cornish, M.L.; Deveau, L.E. Commercialization of Irish moss aquaculture: The Canadian experience. *Bot. Mar.* **2019**, *62*, 411–432. [CrossRef]

92. Liu, J.; Hafting, J.; Critchley, A.T.; Banskota, A.H.; Prithiviraj, B. Components of the cultivated red seaweed *Chondrus crispus* enhance the immune response of *Caenorhabditis elegans* to *Pseudomonas aeruginosa* through the *pmk-1*, *daf-2/daf-16*, and *skn-1* pathways. *Appl. Environ. Microbiol.* **2013**, *79*, 7343–7350. [CrossRef]

93. Liu, J.; Kandasamy, S.; Zhang, J.; Kirby, C.W.; Karakach, T.; Hafting, J.; Critchley, A.T.; Evans, F.; Prithiviraj, B. Prebiotic effects of diet supplemented with the cultivated red seaweed *Chondrus crispus* or with fructo-oligo-saccharide on host immunity, colonic microbiota and gut microbial metabolites. *BMC Complement. Altern. Med.* **2015**, *15*, 279. [CrossRef]

94. Kulshreshtha, G.; Rathgeber, B.; Stratton, G.; Thomas, N.; Evans, F.; Critchley, A.; Hafting, J.; Prithiviraj, B. Feed supplementation with red seaweeds, *Chondrus crispus* and *Sarcodiotheca gaudichaudii*, affects performance, egg quality, and gut microbiota of layer hens. *Poultry Sci.* **2014**, *12*, 2991–3001. [CrossRef] [PubMed]

95. Rudtanatip, T.; Lynch, S.A.; Wongprasert, K.; Culloty, S.C. Assessment of the effects of sulfated polysaccharides extracted from the red seaweed Irish moss *Chondrus crispus* on the immune-stimulant activity in mussels *Mytilus* spp. *Fish Shellfish Immunol.* **2018**, *75*, 284–290. [CrossRef] [PubMed]

96. Johnson, P.J.; Wiedmeyer, C.E.; Messer, N.T.; Ganjam, V.K. Medical implications of obesity in horses—Lessons for human obesity. *J. Diabetes Sci. Technol.* **2009**, *3*, 163–174. [CrossRef] [PubMed]

97. Michalak, I.; Marycz, K. Algae as a promising feed additive for horses. In *Seaweeds as Plant Fertilizer, Agricultural Biostimulants and Animal Fodder*; Pereira, L., Bahcevandziev, K., Joshi, N.H., Eds.; CRC Press: Boca Raton, FL, USA, 2019; pp. 128–142.

J. Mar. Sci. Eng. **2020**, *8*, 574

98. Cordain, L.; Eaton, S.B.; Sebastian, A.; Mann, N.; Lindeberg, S.; Watkins, B.A.; O'Keefe, J.H.; Brand-Miller, J. Origins and evolutionof the Western diet: Health implications for the 21st century. *Am. J. Clin. Nutr.* **2005**, *81*, 341–354. [CrossRef]

99. Armelagos, G.J. Brain evolution, the determinates of food choice, and the omnivore's dilemma. *Crit. Rev. Food Sci. Nutr.* **2014**, *54*, 1330–1341. [CrossRef]

100. Ji, B.W.; Sheth, R.U.; Dixit, P.D.; Tchourine, K.; Vitkup, D. Macroecological dynamics of gut microbiota. *Nat. Microbiol.* **2020**, *5*, 768–775. [CrossRef]

101. Wells, M.L.; Potin, P.; Craigie, J.S.; Raven, J.A.; Merchant, S.S.; Helliwell, K.E.; Smith, A.G.; Camire, M.E.; Brawley, S.H. Algae as nutritional and functional food sources: Revisiting our understanding. *J. Appl. Phycol.* **2017**, *29*, 949–982. [CrossRef]

Journal of
*Marine Science
and Engineering*

Review

Seaweed Potential in the Animal Feed: A Review

Tiago Morais [1], **Ana Inácio** [2,†], **Tiago Coutinho** [2,†], **Mariana Ministro** [3], **João Cotas** [4], **Leonel Pereira** [4] and **Kiril Bahcevandziev** [2,*]

[1]   Lusalgae, Lda, Incubadora de Empresas da Figueira da Foz, Rua das Acácias N° 40-A, 3090-380 Figueira da Foz, Portugal; tsmorais@lusalgae.pt

[2]   Agricultural College of Coimbra (ESAC/IPC), Research Centre for Natural Resources Environment and Society (CERNAS), Institute of Applied Research (IIA), 3045-601 Coimbra, Portugal.; anacmi.31@gmail.com (A.I.); tiago28coutinho@hotmail.com (T.C.)

[3]   Polytechnic Institute of Coimbra/ISEC, Rua Pedro Nunes, Quinta da Nora, 3030-199 Coimbra, Portugal; marianaministro95@gmail.com

[4]   Department of Life Sciences, MARE—Marine and Environmental Sciences Centre, University of Coimbra, 3001-456 Coimbra, Portugal; jcotas@gmail.com (J.C.); leonel.pereira@uc.pt (L.P.)

*   Correspondence: kiril@esac.pt

†   Co-second authors.

Received: 30 June 2020; Accepted: 22 July 2020; Published: 25 July 2020

Abstract: Seaweed (known as marine algae) has a tradition of being part of the animal feed in the coastal areas, from ancient times. Seaweeds, are mixed with animal feed, because when consumed alone can have negative impact on animals. Thus, seaweeds are very rich in useful metabolites (pigments, carotenoids, phlorotannins, polyunsaturated fatty acids, agar, alginate and carrageenan) and minerals (iodine, zinc, sodium, calcium, manganese, iron, selenium), being considered as a natural source of additives that can substitute the antibiotic usage in various animals. In this review, we describe the nutritional values of seaweeds and the seaweed effects in the seaweed-based animal feed/supplements.

Keywords: seaweeds; feed additive; feed supplement; animal nutrition

---

1. Introduction

Seaweeds colonize aquatic habitats and are used mainly by coastal populations [1,2]. Many seaweed species are normally used in unprocessed form, in medicine, human diets, animals feeds and for improvements in agricultural soil as fertilizers [3]. They are rich in potassium, sodium, calcium, magnesium and phosphorus and are a source of essential trace elements, such as iron, manganese, copper, zinc, cobalt, selenium and iodine [4]. Little is known about their bioavailability in nutrients [5]. Seaweeds are simple organisms, which are able to take advantage of sunlight to convert carbon dioxide into sugars and oxygen, during the photosynthesis process. The most common varieties of edible algae include: *Neopyropia/Porphyra/Pyropia* spp., *Undaria pinnatifida*, *Saccharina latissima*, *Palmaria palmata* and *Chondrus crispus*, varieties that are associated with many health benefits, such as decreasing blood pressure, preventing spills and they are a valuable protein source [6].

Seaweed biomass is a valuable alternative ingredient for livestock. Macroalgae, in general, show great differences in proteins, minerals, lipids and fibers. The high mineral content of seaweeds is due to their ability to absorb inorganic substances from the environment; they contain a small amount of lipids, mainly polyunsaturated fatty acids (PUFAs), although they are rich in polysaccharides. Seaweed has only a small percentage of lipids (1–5%), but the majority of those are PUFAs. Predominantly, brown and red seaweed contain more PUFAs 20:5 n-3 (EPA) and 20:4 n-5 (arachidonic acid) than green algae [7–9]. Seaweed has a highly variable composition, which depends on the species, time of collection, habitat and on external conditions such as water, temperature, light intensity and nutrient concentration in water [9,10].

Algae contain high-levels of nonprotein nitrogen, such as free nitrates, resulting in nitrogen-to-protein conversion factors of 5.38, 4.92 and 5.13 for brown, red and green seaweed, respectively [8,9]. There are various edible seaweeds for human consumption with high protein content, with variable essential amino acids [11]. They also absorb minerals from seawater and contain 10 to 20 times more than the land plants [12]. In general, it is accepted that green and red algae have higher nutritional value than brown algae due to low protein and high mineral content. However, brown algae have a higher and diversified content on bioactive molecules with high commercial interest [9]. Therefore, algae can provide energy, minerals and proteins to animal feed and have potential as alternative protein source for ruminants [13].

The analysis of the protein quality and concentration is essential to determine the nutritional value of the algae biomass, so it can be used fresh, dried, liquefied or cooked. This analysis is important as it identifies the concentration of essential amino acids (EAAs) [14].

Algae have a relatively high protein quality compared to cereal and soy flour. More than 75% of seaweed has higher proportions of total essential amino acids than wheat flour and 50% higher than soy flour and also higher than rice and corn [14–16]. The proportion of EAAs methionine and lysine are comparable to traditional protein sources. Algae are generally richer than soy flour in the proportion of methionine but poorer when compared to wheat flour. On the other hand, algae have a lower proportion of lysine than soy and wheat flour [14].

The main limitation for the use of seaweed proteins is the concentration of EAAs and not the quality of the total amino acids or proteins. However, the concentration of EAAs in seaweed, in an entire biomass base, is considerably lower than in traditional sources, such as corn and soy, thus it is not suitable as a protein source in compound diets for monogastric animals. This does not detract from its positive health benefits for humans and livestock, where its few calories and high mineral content may be desirable [14].

Among the marine organisms, seaweed represents one of the richest sources of natural antioxidants and antimicrobials. They are also an excellent source of vitamins such as A, Bl, B12, C, D and E, riboflavin, niacin, pantothanic acid, folic acid as well as minerals such as Ca, P, Na, K and I [17].

In short, new alternatives to reduce or replace the use of antibiotics in animal diet is needed. Thus, we can contribute to find a natural product that does not only eliminate or prevent diseases but also improves the nutrient quality of meat and eggs. With the continued study of the sea resources, numerous species of algae with favorable biological activity have been reported as acceptable for inclusion in diets for rats, chickens, laying hens and pigs [18].

In this review, our intention is to analyze the seaweed potential for animal feed and to contribute to the development of its standardized use, reducing the animal health risks.

2. Seaweeds: Nutritional Profile

As demonstrated earlier, seaweeds (or macroalgae) are divided into three large groups, without any taxonomic value, based on the color they present [19,20]. There are 10,100 seaweed species known worldwide and they can be observed in all seawater habitats with some seaweeds appearing in freshwater [21].

Green seaweeds (Chlorophyta), of which there are known to be 2200 species, at maximum reach 1 m in height. Their color is related to the presence of chlorophyll [4,21,22]. Red seaweeds (Rhodophyta), with 6100 species, are efficient in photosynthesizing in deeper waters. Their length varies and they are similar to green seaweeds. Their color results from the presence of pigments, phycoerythrin and phycocyanin, which disguise β-carotene, lutein, zeaxanthin and chlorophyll [4,21,23–25]. For brown seaweeds (Ochrophyta, Phaeophyceae), with nearly 1800 species, only 1% is known to exist in fresh water and their length can be up to 50 m. The brown color is related to their content of carotenoid fucoxanthin, which disguises β-carotene, violaxanthin, diatoxanthin and chlorophyll. The main seaweed polysaccharides are laminarin, fucoidans, agar, carrageenans, porphyran, ulvans and alginates [4,21,26,27].

Seaweed mainly contains high levels of glutamic acid, present in both free and protein-bound forms, contributing to typical flavor known as umami. They also contain various bioactive amino acids and peptides, such as taurine, carnosine and glutathione [5]. In this point we intend to address the nutritional value and other relevant molecules found in the different seaweeds.

2.1. Green Seaweeds

The commonly known green algae are organisms which belong to the Chlorophyceae class (phylum Chlorophyceae), including both microscopic and macroscopic species. They are the most diverse algae group, with more than 13,000 species; it is estimated that about half of these species are seaweeds. The characteristic color is due to the presence of the chlorophyll *a* and *b*, pigment used during the photosynthetic process [2,28].

Their color usually depends on the balance between these chlorophylls and other pigments, such as β-carotene and xanthophylls. Green algae are common in areas where light is abundant, such as shallow water and natural pools. The main genus include *Ulva*, *Codium*, *Chaetomorpha* and *Cladophora* [4].

Ulva is a one of the most common genera of green seaweeds, also found in brackish water (mainly in estuaries). Being filled with minerals, proteins and vitamins, these species are very appealing to study at a nutritional level [3]. *Ulva*'s biomass is relatively rich in proteins (Table 1) and has a potential as an alternative source of proteins for animal feeding, contains highly insoluble dietary fibers (glucans) and soluble fibers, having higher protein content than other green seaweeds [4].

Ulva sp. grows abundantly in areas rich in nutrients, float together along the coast, block watercourses and destroy the marine ecosystem, which becomes a serious threat to the fishing industry and the development of tourism. Consequently, it could be of great practical interest to make *Ulva* waste profitable. This seaweed is not studied only for its high protein levels. The interest, both academic as commercial, in the abundant and highly sulfated ulvans that are extracted from *Ulva*, has increased in recent years. Ulvan is a heteropolysaccharide of the cell wall that represents 9 to 36% in dry weight of *Ulva* sp. biomass [29]. Ulvan consists of rhamnose, xylose, glucose, uranic acid and sulfate, which regulate immune functions and act as an antioxidant and antibiotic [22]. A high level of this sulfated polysaccharide in *Ulva* sp. reveals its anticoagulant, antiviral, anti-inflammatory, antihyperlipidemic, immunomodulatory and anticancer activities [18].

Some *Ulva* species are used as livestock feed [30] and adding *Ulva* to diets in powder form can decrease an abdominal and subcutaneous fat, improving meat quality and amylase activity in the duodenal content of chicken [5].

The use of *Ulva* as a food or animal feed is a daunting task since the bioavailability of the polysaccharides (Table 1) has remained indescribable due to the inefficient animal metabolism regarding this nutrient. This chemical limitation generally prevents the efficient use of *Ulva* as a single feed for animals [3].

Table 1. Nutrient (% dry weight) and mineral (mg 100 g^{-1} dry weight) composition of some edible seaweeds [31].

Species	Nutrient Composition (%)					Mineral Composition (mg.100 g^{-1})				
	Protein	Ash	Dietary Fiber	Carbohydrate	Lipid	Na	K	P	Ca	Mg
Green seaweed										
Caulerpa lentillifera	10–13	24–37	33	38–59	0.86–1.11	8917	700–1142	1030	780–1874	630–1650
C. racemosa	17.8–18.4	7–19	64.9	33–41	9.8	2574	318	29.71	1852	384–1610
Codium fragile	8–11	21–39	5.1	39–67	0.5–1.5	-	-	-	-	-
Ulva compressa	21–32	17–19	29–45	48.2	0.3–4.2	-	-	-	-	-
U. lactuca	10–25	12.9	29–55	36–43	0.6–1.6	-	-	140	840	-
U. pertusa	20–26	-	-	47.0	-	-	-	-	-	-
U. rigida	18–19	28.6	38–41	43–56	0.9–2.0	1595	1561	210	524	2094
U. reticulta	17–20	-	65.7	50–58	1.7–2.3	-	-	-	-	-
Red seaweed										
Chondrus crispus	11–21	21	10–34	55–68	1.0–3.0	1200–4270	1350–3184	135	420–1120	600–732
Crassiphycus changii	6.9	22.7	24.7	-	3.3	5465	3417	-	402	565
Agarophyton chilense	13.7	18.9	-	66.1	1.3					
Palmaria palmata	8–35	12–37	29–46	46–56	0.7–3	1600–2500	7000–9000	235	560–1200	170–610
Neopyropia teneraNeopyropia tenera	28–47	8–21	12–35	44.3	0.7–1.3	3627	3500	-	390	565
Porphyra umbilicalis	29–39	12	29–35	43	0.3	940	2030	235	330	370
Neopyropia yezoensis	31–44	7.8	30–59	44.4	2.1	570	2400	-	440	650
Brown seaweed										
Alaria esculenta	9–20	-	42.86	46–51	1–2	-	-	-	-	-
Eisenia bicyclis	7.5	9.72	10–75	60.6	0.1	-	-	-	-	-
Fucus spiralis	10.77	-	63.88	-	-	-	-	-	-	-
F. vesiculosus	3–14	14–30	45–59	46.8	1.9	2450–5469	2500–4322	315	725–938	670–994
Himanthalia elongata	5–15	27–36	33–37	44–61	0.5–1.1	4100	8250	240	720	435
Laminaria digitata	8–15	38	37–37	48	1.0	3818	11.5–79	-	1005	659
L. ochroleuca	7.49	29.47	-	-	0.92	-	-	-	-	-
Saccharina japonica	7–8	27–33	10–41	51.9	1.0–1.9	2532–3260	4350–5951	150–300	225–910	550–757
S. latissima	6–6.26	34.78	30	52–61	0.5–1.1	2620	4330	165	810	715
Sargassum fusiforme	11.6	19.77	17–69	30.6	1.4	-	-	-	1860	687
Undaria pinnatifida	12–23	26–40	16–51	45–51	1.05–45	1600–7000	5500–6810	235–450	680–1380	405–680

2.2. Red Seaweeds

In general, compared to green and brown algae, red algae contains a high amount of proteins (Table 1) reaching 47% (*Neopyropia tenera*) of a dry matter [32].

The proteins from this seaweed group are made up of one or more chains of amino acids, especially glycine (Gly), alanine (Ala), arginine (Arg), proline (Pro), glutamic (Glu) and aspartic (Asp) acid (compose large part of the amino acid fraction), whereas tyrosine, methionine and cysteine appear in a lower quantity. Glutamic and aspartic acid, that have acidic side chains at neutral pH, in red seaweeds represent 14–19% of amino acids [33]. Dawczinski et al. [11] found relevant values of the amino acid taurine (tau) in red seaweeds unlike the brown seaweeds. Essential amino acids reveal almost half of the total amino acids and their protein profile is close to the egg's protein profile. In general, all algae have the same amount of nonessential amino acids [11].

Lipids, in these seaweeds present relatively lower contents, 1–5% of dry matter (Table 1), found in *Chondrus crispus* (1.0–1.3%) and *Palmaria palmata* (0.7–3%) [27,34]. Red seaweed predominantly contains the polyunsaturated 20 carbon-fatty acids eicosapentaenoic acid (EPA, ω-3, C20:5) and arachidonic acid (AA, ω-6, C20:4). Palmitic acid (C16:0) is the main saturated fatty acid with 26% and monounsaturated is oleic acid [11]. *Neopyropia/Porphyra/Pyropia* sp. were tested and the assays showed that palmitic, eicosapentaenoic, arachidonic, oleic, linoleic and linolenic acid were the main fatty acids [11]. Another class of lipids are sterol compounds. Most red algae contain cholesterol, desmosterol, sitosterol, fucosterol and chalinasterol [35].

In their study, Dawczinski et al. [11] did not find significant differences between the red and brown algae, as they both revealed low fat and high fiber content (Table 1). Red algae contains soluble fibers such as sulphated galactans (agars and carrageenans), xylans and floridean starch [11,35,36].

Red algae (Rhodophyta) are seaweeds with an interesting nutritional profile. The minerals present in some red algae, namely *Chondrus crispus* and *Gracilariopsis* sp., are Na, K, Ca and Mg, as well as, Fe, Zn, Mn and Cu [36]. The iodine content in red algae is high, in *Gracilaria* sp. reaching 426 mg/100 g of seaweed dry biomass but not as high as in brown algae (1200 mg/100 g of seaweed dry biomass). Algae iodine has already contributed to the nutritional enrichment of the meat of several fish species [37].

Red seaweed outperformed several brown and green seaweeds in sequestering negatively charged hexavalent chromium ions. It possesses more cationic sites, which show low affinity for positively charged metal ions, such as cadmium, but higher affinity for hexavalent chromium [10].

Algae may have bioactive compounds and bioactive secondary metabolites. Red algae are the main source of halogenated monoterpenes. Seaweed contains also sesquiterpenes, diterpenes, C15-Acetogenins, C27 and C28 steroids, with C29 steroids, in small amounts [38].

Most red seaweeds contain water-soluble vitamins B and C (Table 2), mainly amine and riboflavin and liposoluble vitamins such as carotenoids (as provitamins of vitamin A). The carotenoids are represented by different pigments which form the resulting seaweed color together with chlorophyll and are also very strong antioxidants. The main carotenoids of red seaweed are α-and β-carotene and their derivates such as zeaxanthin and lutein [10,36].

Table 2. Vitamins contents (mg 100 g^{-1} edible portion) of seaweeds [31].

Species	A	B$_1$	B$_2$	B$_3$	B$_5$	B$_6$	B$_8$	C	E	Fatty Acids
Green seaweed										
Caulerpa lentillifera	-	0.05	0.02	1.09	-	-	-	1.00	2.22	-
C. racemosa	-	-	-	-	-	-	-	-	-	-
Codium fragile	0.527	0.223	0.559	-	-	-	-	<0.223	-	-
Ulva compressa	-	-	-	-	-	-	-	-	-	-
U. lactuca	0.017	<0.024	0.533	98 *	-	6 *	-	<0.242	-	-
U. pertusa	-	-	-	-	-	-	-	30–241 **	-	-
U. rigida	9581	0.47	0.199	<0.5	1.70	<0.1	0.012	9.42	19.70	0.108
U. reticulta	-	-	-	-	-	-	-	-	-	-
Red seaweed										
Chondrus crispus	-	-	-	-	-	-	-	10–13 *	-	-
Crassiphycus changii Agarophyton chilense	-	-	-	-	-	-	-	16–149 **	-	-
Palmaria palmata	1.59	0.073–1.56	0.51–1.91	1.89	-	8.99	-	6.34.34.5	2.2–13.9	0.267
Neopyropia tenera	-	-	-	-	-	-	-	-	-	-
Porphyra umbilicalis	3.65	0.144	0.36	-	-	-	-	4.214	-	0.363
Neopyropia yezoensis	16,000 ***	0.129	0.382	11.0	-	-	-	-	-	-
Brown seaweed										
Alaria esculenta	-	-	0.3–1 *	5 *	-	0.1 *	-	100–500 *	-	-
Eisenia bicyclis	-	-	-	-	-	-	-	-	-	-
Fucus spiralis	-	-	-	-	-	-	-	-	-	-
F. vesiculosus	0.30–7	0.02	0.035	-	-	-	-	14.124	-	-
Himanthalia elongata	0.079	0.020	0.020	-	-	-	-	28.56	-	0.176–0.258
Laminaria digitata	-	1.250	0.138	61.2	-	6.41	6.41	35.5	3.43	-
L. ochroleuca	0.042	0.058	0.212	-	-	-	-	0.353	-	0.479
Saccharina japonica	0.48	0.2	0.85	1.58	-	0.09	-	-	-	-
S. latissimi	0.04	0.05	0.21	-	-	-	-	0.35	1.6	-
Sargassum fusiforme	-	-	-	-	-	-	-	-	-	-
Undaria pinnatifida	0.04–0.22	0.17–0.30	0.23–1.4	2.56	-	0.18	-	5.29	1.4–2.5	0.479

* expressed as ppm; ** expressed as mg%; *** expressed as I.

2.3. Brown Seaweeds

In general, brown algae (Phaeophyceae) are seaweeds with the lowest protein content (Table 1), when compared to red and green algae. The most frequently determined protein content in brown seaweeds occurs within a declared range of 5 to 15% [9,10].

The concentrations of EAA in brown algae differ substantially between species. The concentrations of threonine (Thr), valine (Val), isoleucine (Ile), leucine (Leu), phenylalanine (Phe), lysine (Lys) and methionine (Met) were higher in *Undaria pinnatifida* than in *Laminaria* sp.; however, *Laminaria* sp. had higher concentrations of cysteine (Cys) than *Undaria pinnatifida*. Aspartic acid and glutamic acid amino acids were the most abundant in these algae species tested in this study [10]. Brown algae contained higher concentrations of phosphoserine than red algae [10] while glutamic and aspartic acid represents 20–44% [32].

The type of carbohydrate varies greatly among the algae. The soluble fibers are alginates, fucans and laminarins for brown seaweeds. Fucoidans, sulphated polysaccharides, are extensively involved in the cell walls of brown seaweed [29]. In terms of dietary fiber (Table 1), it is not uniform in all brown algae [10]. Fucoidans present several physiological and biological characteristics, such as antitumor, anticoagulants, antioxidant, antiviral and antithrombotic activities, besides the impact on the inflammatory and immunological systems [28].

According to some researchers, laminarin is the second main source of glucan in brown algae and it was detected as a regulator of intestinal metabolism through its impact on mucus structure, intestinal pH and short chain fatty acid formation [29].

Brown algae are balanced sources of omega-3 and omega-6 acids [29,36]. The brown seaweeds, such as *Undaria pinnatifida*, *Laminaria* sp. and *Hizikia fusiforme*, contain predominantly 20 polyunsaturated eicosapentaenoic acids (EPA, ω-3, C 20:5) and arachidonic acid (AA, ω-6, C 20:4) [10]. Palmitic acid (C16:0) is one of the most abundant but not as abundant as in most red algae. In general, other fatty acids, abundant in brown algae, are the essential fatty acids, oleic acid (ω-9, C18:1), linoleic acid (ω-6,C18:2), linolenic acid (ω-3,C18:3) and the precursors of the eicosanoids, arachidonic acid (ω-6,C20:4) and eicosapentaenoic acid (ω-3,C20:5) [36]. Some saturated and monounsaturated fatty acids are found in abundance (Table 2) only in some brown algae (*Laminaria* sp. and *Undaria pinnatifida*) such as myristic and palmitolenic acid [10].

Polyphenols (fucol, fucophlorethol, fucodiphloroetol G ergosterol) and the phenolic compound phlorotannin are also abundant, in *Sargassum*, *Fucus* and *Ascophyllum nodosum*, and they have strong antioxidant effects. These seaweeds also contain halogenated compounds [31].

The minerals present in some brown algae (Phaeophyceae), namely *Undaria pinnatifida* (Table 1) and *Sargassum* sp., are some of the main ones (Na, K, Ca and Mg), as well as Fe, Zn, Mn and Cu [29]. In certain brown algae the concentration of iodine can reach very high levels, in particular the genus *Laminaria*. According to several authors, *Saccharina japonica* (as *Laminaria japonica*) presented the highest iodine content of 5.6 and 3.04 mg/kg among other seaweeds studied [9,31].

Brown algae can participate in the accumulation of metals due to their carboxyl groups and because the cell wall is formed by cellulose. They have cationic characteristics but less than red algae [9].

Brown algae contain considerably higher concentrations of arsenic than red or green algae. In most species of seaweeds, the concentrations are below 54 mg kg^{-1} dry weight, and 5–10% of the total arsenic is organic [38].

Some of the most important vitamins present in most brown algae are vitamin C, E and group B vitamins (Table 2), especially thiamine and riboflavin [9]. Vitamin B12 is present in brown algae in lower concentrations than red and green algae [29]. Brown seaweed contains larger amounts of vitamin E and high amounts of vitamin C. Brown seaweed carotenoids are formed by fucoxanthin, β carotene, lutein, violaxanthin, antheraxanthin, zeaxanthin and neoxanthin. Fucoxanthin is the main carotenoid in brown seaweed and has been shown to have many health benefits [9].

3. Seaweed as Valuable Nutritional and Nutraceutical Animal Feed

The nutritional value attributed to macroalgae along with their nonanimal nature makes them particularly appropriate to be used in animal feed as nutraceuticals, a term that results from the combination of nutritional and pharmaceutical, used to identify food components that bring health benefits, including the prevention to some diseases [39,40].

The health benefits of seaweed, beyond the provision of essential nutrients, have been supported by in vitro studies and some animal studies; however, many of these studies have inappropriate biomarkers to substantiate a claim and have not progressed to suitably designed trials to evaluate efficacy. The limited evidence that does exist makes some seaweed components attractive as functional food ingredients, but more animal nutritional studies evidence (including mechanistic evidence) is needed to evaluate both the nutritional benefit conferred and the efficacy of purported bioactivities and to determine any potential adverse effects [41]. Through an evaluation of the nutritional composition of edible seaweeds in Section 2, this section summarizes the available evidence and outlines the potential of the seaweeds as animal feed hypothesis with a prominent feed safety question.

3.1. Feed Safety

The animal feed plays a vital role in the global food security, and it is conceived to ensure the sustainable production of safe and affordable animal proteins. With the increase of the animal production, it will be necessary for more feed to be produced, which will be safety certified. Consequently, new and old feed sources are being controlled for hazards and critically analyzed to guarantee feed safety for animal consumption. However, the food safety regulatory framework is not fully harmonized between the countries, creating a lack in feed safety chain, increasing the animal health risks and the animal consumption by the humans [42].

Seaweed are considered a rich and sustainable source of macronutrients (particularly dietary fiber) and micronutrients to the animal feed, but if seaweeds are to contribute to future global food security, legislative measures to ensure monitoring and labeling of feed products are needed to safeguard against excessive intakes of salt, iodine, and heavy metals, such as arsenic (As), aluminum (Al), cadmium (Cd), lead (Pb), rubidium (Rb), silicon (Si), strontium (Sr) and tin (Sn) [43,44]. While heavy metal concentrations in seaweeds are generally below toxic levels, bioaccumulation of arsenic and lead are the main risk in wild seaweed harvest, and more studies of heavy metal toxicokinetics are needed to address the problem.

Levels of arsenic, mercury, lead, and cadmium in 426 Korean dried seaweed products ranged from 0.2 to 6.7% of provisional tolerable weekly intakes when 8.5 g of seaweed was consumed per day in human food consumption [41]. Chen et al. [43] revealed the different levels of Al, As, Cd, Cr, Cu, Hg, Mn, Ni, Pb and Se in dried seaweeds from southeastern China (Zhejiang province). This indicates that element concentration changes with different species of seaweeds and origin areas. For example, the levels of Cd, Cu, Mn and Ni in red seaweeds (*Porphyra*) were significantly higher than those in brown seaweeds (*Laminaria*, *Saccharina* and *Undaria*).

A tradeoff between iodine and/or heavy metal ingestion and the amount of whole seaweed needed to obtain meaningful amounts of PUFAs, protein or dietary fiber may limit the recommended portion size of the seaweeds concentration in feed [41]. Relevant and key information to use seaweeds with feed safety guarantee will be gathered. However, for most countries there is no regulation on maximum levels of heavy metals in seaweed [43]. However, there is Regulation (EC) No 1831/2003 laying down rules governing the European Community authorization of feed additives. In addition, Regulation (EC) No 429/2008 lays down detailed rules for the implementation of Regulation (EC) No 1831/2003 as regards the preparation and the submission of applications and the assessment and the authorization of feed additives [45].

The production of rancid flavors and odors due to oxidative stress can lead to a reduction in the sensory attributes, nutritional quality and food safety. Extracts from seaweeds are rich in polyphenolic compounds which have well documented antioxidant properties. They also have antimicrobial activities against major food spoilage and feed pathogenic micro-organisms. The addition of seaweeds or their extracts to feed products will reduce the utilization of chemical preservatives, which will fulfill the industry as well as consumer demands for "green" products. In addition, the current status and the future projections in the functional effects of seaweeds as a means to improve the fiber content and reduce the salt content of food products will be of significant importance to the meat industry [17].

3.2. Fish Farming

Over 50% of the operating costs in an intensive fish aquaculture are related to the fish feeding [46,47]. This led to research, in recent decades, for new sources of aquaculture nutrients, especially on terrestrial plants as legumes and oilseed crops [47–49]. Algae are a natural alternative to soybean for fish diets, presenting economic and nutritional advantages, since the nutritional profiles made to soybean show that this plant does not fully match the fish nutritional requirements [47,50,51]. This problem has been attributed to using a certain amount of plant protein sources which might contain antinutritional factors and result in palatability problems [47,52,53].

In recent years, there were some experiments carried out with the objective to find more economically and nutritional viable options for fish feed. Al-Hafedh et al. [54] experimented with the application of green *Ulva* and red seaweeds *Gracilaria* with the objective to reduce nutrient concentrations in sea water effluent, sources of pollution and to diversify origin of the feeds in the changing market status as offering the possibility for additional sources of income. This was considered to be highly relevant to develop the industry of aquaculture [55] and to reduce the dangers of an oligotrophic sea that has a high level of biodiversity [47,56,57]. The Integrated Multitrophic Aquaculture (IMTA) is a cultivation system that is based in the fish or shrimp (herbivores or carnivores) aquaculture with cultivation of others species, such as sea urchins or bivalves and seaweeds to capture the excess of nutrients from the effluents of the fish or shrimp tanks or cages. Basically, the seaweeds (inorganic nutrients absorber) "clean" the nutrient-enriched water output from fish or shrimp aquaculture, thus providing a possible source for the aquatic animal feed. The IMTA provides sustainable conditions to an intensive culture of fish or shrimp, which is being practiced in many countries, as integrated unit with seaweeds and mollusk culture. This approach, besides being a form of balanced ecosystem management, prevents potential environmental impacts from aquaculture. It also provides exciting new opportunities for valuable crops of seaweeds, transforming it in one of the fields for further technological advances in seaweed aquaculture [58].

Other experiences with diverse seaweeds, such as *Ulva* sp., *Neopyropia/Porphyra/Pyropia* sp., *Gracilaria* sp., *Ascophyllum nodosum*, *Sargassum* sp. and *Padina* sp. led to encouraging results for the use of seaweed as fish feed, which appears to depend on seaweed species, its incorporation level and the fish species where the seaweed is produced [59–68]. In short, it was discovered that using seaweed meals as supplement to fish diets enhance the growth, lipid metabolism, physiological activity, stress response, disease resistance and carcass quality of various fish species [69–71].

In 2019, Kamunde et al. [72] studied a salmon meal based on brown seaweed *Laminaria* sp. named AquaArom®. The protocol was intended to access food intake, growth performance, plasma antioxidant capacity and mitochondrial respiration. The conclusions showed that the addition of AquaArom® to commercial salmonid food improves those characteristics and alleviates the effect of temperature rise on mitochondrial respiration. The slight decline in crude protein and minerals resulting from the addition of up to 10% AquaArom® to aquafeed appear to have no adverse consequences on Atlantic salmon smolts. Thus, mixing of brown seaweed meal with commercial aquafeeds (and potentially feeds from other farm animals) could offer a cost-effective way of harnessing the beneficial effects of seaweeds in animal production. The lower level of protein accessed in this experiment, resulted from the use of brown seaweed (known for low crude protein levels), and it may be complemented by green or red seaweed [72]. Other brown gigantic seaweeds (commonly, named kelp), *Saccharina latissima* (Phaeophyceae) demonstrated potential to be inserted as feed additive (with concentration below 4%) for rainbow trout, where the fatty acids (e.g., oleic and linoleic acids) and lipid fraction in the fillet was reduced, although the eicosapentaenoic acid and docosahexaenoic acid (omega-3 PUFAs) was not affected. Kelp supplementation also increased a protective activity against oxidative stress in this fish [73], being in line with the data gathered about kelp, which points out that, as a brown seaweed, it produces secondary metabolites with more acceptable nutraceutical characteristics.

In conclusion, there is a great quantity of data to support the use of seaweed in order to achieve a higher productivity on fish farms. The advantages range from growth and development rates, disease resistance, financial gain and even ecological preservation. This is a case where the implementation of such research and development (R&D) was a relatively easy process, leading to various companies developing their own products, as exemplified earlier with AquaArom®, proving, once again, the safety of using seaweed as fish feed.

3.3. Oyster Feed

Oyster is a highly valorized and appreciated seafood product. It is one of the most widely cultivated marine animals that in 2014 exceeded 600,000 tones worldwide [55].

Production of oysters in hatchery encompasses three distinct stages: broodstock conditioning, larval culturing and postlarval rearing phase. Broodstock conditioning is a stage of the utmost importance in hatchery because it is the first chance to modulate/condition the whole offspring. A higher oyster fecundity, better quality eggs and enhanced larval viability are possible through intelligent innovations in broodstock conditioning [74,75]. It has been shown that physical and nutritional factors can modulate gonadal development, either accelerating gametogenesis or slowing gonadal maturation [76]. Several nutritional studies were made on bivalve species that have tested different algal compositions for an enhanced reproductive outcome [74,77–80]. Currently, the most successful strategy to a balanced diet lies in microalga blends; due to its nutritional value, it is possible to obtain optimal food conditions [81]. This entails a dependence on the production of live microalgae, which may represent up to 30–40% of the hatchery operational costs [82] but can show an economic limitation for the use of such a diet in bivalve hatcheries [83,84]. This leads to a new research line focused on seaweed—it can be a nutritional source for oyster, human and animals [27].

Nutritional profile influences the physiology of bivalves, having a strong effect on their proteins, carbohydrates and lipids [85,86], being that these organisms are known to be mineral accumulators. For example, bivalves are rich in potassium, sodium, calcium, magnesium, phosphorus and are a source of essential trace elements, such as iron, manganese, copper, zinc, cobalt, selenium and iodine [87]. Some species of green seaweed, such as *Ulva* sp., have high protein contents (10–25% in dry content) and high levels of mineral elements with nutritional value, including calcium and magnesium [88]. This suggests the hypothesis of seaweed use as feed, at least partially, on oyster and other bivalve feeds; moreover, they have been shown to improve stress response and resistance to disease, thereby representing a meaningful advantage to aquaculture [88].

Before further advancing, there is a fundamental question that has to be addressed since seaweeds (for example *Ulva* sp. *and Fucus* sp.) are also known to be mineral element accumulators, some of them highly toxic for humans, for example iodine, arsenic and mercury [89]. As seaweed can accumulate highly toxic minerals, so can oysters, using seaweeds as food, that can represent a way to introduce hazardous elements into the oyster feeding and, respectively, human food. Regarding this, in order to evaluate the potential nutritional value of oysters, it is crucial to know whether any given nutrient, as part of feed, can or cannot be bio-accessible for humans [89,90].

Cardoso et al. [90] designed an experimental protocol to determine better macro and microalgae blend to feed pacific oyster (*Crassostrea gigas*) that describes the bio-accessibility of nutrients and minerals. It was observed that oysters consuming only one seaweed species, independent from species studied, had the highest levels of Be, Cu, Zn, Sr and Cd. The most important problem in the oysters' composition is the increment of microalgae concentrations in the oysters' feeding system with a progressive concentration reduction of seaweed. When high levels of Cd or Pb were found or Zn in oysters, the study indicates that caution and further study are needed to guarantee and maintain low the heavy metal levels during the substitution of the feed source. It was also observed that Mn, Cd and Pb bio-accessibility has increased with the substitution of the initial microalgal with seaweed feed, proving the seaweed potential as the oyster feed with reduction of the above cited dangerous elements [90].

As demonstrated in this topic, there is promising data to support the use of green seaweed, mainly *Ulva* sp. in order to achieve a higher productivity of oyster farms due to high protein contents and high levels of mineral elements found in oysters. These advantages can also result from improved stress response, resistance to disease, financial gain and ecological preservation. However, a capacity of metal accumulation, some of them toxic, shared between seaweed and oysters, raise some questions, which must be taken into account during the development of the oyster feed product.

3.4. Poultry Feeds

When addressing the issue of poultry farming, it is necessary to distinguish between poultry raised for the consumption of their meat (broiler poultry) and poultry raised for egg consumption (laying poultry). Thus, subjects will be addressed in two different points.

There are three main reasons to use seaweed in poultry feed: to improve animal immune status, to decrease microbial load in the digestive tract and for beneficial effects on the meat and eggs [2,89,91–95].

3.4.1. Broiler Poultry

In broiler aviaries, feed is based primarily on corn and soybean meals, with corn in most parts of the world as a source of energy due to its abundance and digestibility (60–75% of broilers diet). Historically, high corn prices led to the search for new feed capable to provide the required nutrients for broilers, in order to maintain productivity and lower the feed price.

Green algae (Chlorophyta) have been studied in the previous century as an alternative to feed poultry. Asar [96] found that supplementation of chicken's basal diet with 4% of seaweed increased body weight gain. El-Deek et al. [97] found no significant effects on growth, feed intake and feed conversion ratio with the inclusion of *Sargassum* ssp. (Phaeophyceae) on broiler diet. Gu et al. [98] concluded that a 2% seaweed inclusion on the broiler feed improved performance and dressing percentage. Ventura et al. [99] compared animal feeds with two different concentrations of *Ulva rigida* (10 and 20%); the data obtained indicated that the feed intake and body weight gain was better with 10% of *U. rigida*. The feed with 20% of *U. rigida* had a harmful effect on the broilers. Later, it was found that poultry fed with 10% mixture of green algae, containing various species as genus *Ulva*, *Caulerpa*, *Codium*, *Halimeda* and *Bryopsis*, showed better growth with statistical differences in body weight, a lower level of fats (0.7–1.7%) and higher protein contents (46.6–72.2%) when compared to control groups (1.1–3.2% and 66.4–71.4%, respectively). Species like *Codium* sp., with spongy thallus, can retain high amounts of salt in this structure which can lead broilers to lose weight because of diarrhea [100].

Studies with another *Ulva* species, *U. lactuca* (Sea lettuce), which can be found on Atlantic shores, pointed out that, with lower than 3% seaweed added to animal diet, broilers performed better than the respective control diet. It was speculated that higher crude protein and amino acids, especially methionine, plays an important role in the improvement of dressing and breast yield. However, as mentioned before, seasonal variations in nutrient composition of seaweed must be considered [93]. With this result, *U. lactuca* can be pointed out as a more economical ingredient to be incorporated, at least partially, in broiler feeding.

The nutraceutical characteristics of seaweed have been the subject of studies in recent years. The work of Kulshreshtha et al. [101] included red seaweeds (*Chondrus crispus* and *Sarcodiotheca gaudichaudii*) in livestock feed with the objective to study its nutritional value and the prebiotic potential. The research started with the fact that antibiotics are used to stimulate growth and to control disease-causing pathogens in layer chickens [101–104]. The prolonged and indiscriminate use of antibiotics in livestock led to concerns such as development of antibiotic-resistant strains of pathogens, high concentrations of antibiotic residues in meat and meat products and undesirable changes in the microbial communities of animal gastrointestinal tracts [101,105–107] and as a result, numerous countries, including the UE, banned the use of antibiotics as a growth promoter. There was not significant improvement found by joining red seaweed in broiler feed in parameters as growth, development, feed intake or egg production, when compared to control [101]. However, the inclusion of red seaweed is more interesting from the prebiotic point of view. In fact, this research found that the seaweed species included in the meal showed an increase in the population of beneficial bacteria and a reduction of pathogenic bacteria in the gut, improvement in villi height, crypt depth, and an increase in the concentration of short chain fatty acids, which can also be replicated in laying hens [101,108]. These results are directly linked to the fact that red algae contain specific bioactive compounds, such as agars, carrageenans, xylans, sulphated galactans and porphyrins that may be responsible for the effects [109–111]. However, further study is needed to determine such mechanisms.

As a practical conclusion, and based in what was described, it appears to be possible to enrich broiler feed with green seaweed, or a mixture of green and red seaweed, in order to stimulate both the growth and the health of the broilers. The limiting factor appears to be the use of low concentration (1–2%) of seaweed in the meal, which could represent a healthier method to achieve the proposed objectives, as well as, a less expensive one, when compared to the actual methods [111]. Following this data, the next logical step should be the investment in R&D work in order to create products, based on seaweeds, able to be included in the market as an alternative to the existing ones.

3.4.2. Laying Poultry

Eggs are one of nature's most wholesome foods because of their content in essential and nonessential minerals, high-quality proteins, lipids and vitamins. Egg composition can be altered by hereditary genes, diet and poultry age. Egg yolk contains natural carotenoids, and its yellow color is attributed to β-carotene, zeaxanthin, kryptoxanthin and lutein, which are easily found in commercial feed [39,112–114]. Alongside those carotenoids, eggs represent an important source of protein, minerals (phosphorus, iron) and easily digestible fats (93–96%) like ω-3 fatty acids, all of which can be enriched by supplementation of the poultry feed [40].

There was research in the last decade with the objective to enrich the egg molecular content and to adjust it for better human consumption. This can be achieved by supplementation of the poultry feed with seaweed, which can be used to enhance the levels of vitamins, minerals and fatty acids, mainly ω-3 fatty acids [115–117].

Research was done with green algae from the genus *Ulva*, with the inclusion of 1–3% of this seaweed resulting in improved egg production and quality, increasing the weight, shell thickness, yolk color and reduced yolk cholesterol. The seaweed extract also reduced *Escherichia coli* load in feces, which suggests better health of the animals and a decreasing feed conversion ratio [2,95]. These results need further studies in order to access the bioavailability of seaweed contents in order to determine what concentration is the best.

Recently, red seaweed, such as *Chondrus crispus*, has been used at 2–4% feed to reduce the level of *Salmonella enteritidis*, a toxic bacterium which can be transmitted vertically from laying hens to eggs through the ovaries and oviducts or due to contaminated feces. Reducing the level of this bacterium is vital to produce safer eggs and reduce the spread of salmonellosis to humans. A greater reduction of negative effects was observed on layer growth and egg production caused by *S. enteritidis* with this seaweed at the 4% diet. Dietary inclusion of *C. crispus* inhibited colonization of Salmonella in the excreta and ceca, which could be due to promoting the growth of *Lactobacillus* and increasing the concentration of short chain fatty acids. A higher level of IgA in birds supplemented with this feed indicates a direct role of seaweed on the maturation of the humoral immune system. These results encourage the research for a nontoxic alternative feed that producers (including organic farmers) can accept [108].

Using brown seaweed has the potential, like *Sargassum* sp., at a 3–6% dietary level, to give benefits to the egg quality, decreasing yolk cholesterol, triglycerides and ω-6 fatty acids and increased carotene and lutein plus zeaxanthin contents. There are data of poultry being fed with boiled seaweed which resulted in improvement of the high density lipoprotein, which is beneficial for human health [118]. The research even approached the way in which seaweed should be used as feed to layers, designing a protocol in which groups should be applied as sundried, boiled or autoclaved seaweed at 3 and 6%. This approach intends to expand further on which edible brown seaweed offers a variety of health benefits, mainly due to the relatively high contents of ω-3, Ca and Fe. Results have shown that there are differences between the groups. However, it was concluded that those differences could be due to geographical location, year season, environmental factors, growth media and physiological conditions [118]. Further research should include controlled production of seaweeds (in aquaculture, for example) in order to maintain a seaweed stability profile and minimize the influence of such external factors. In this way it should be possible to determine how laying poultry feed should be administered in order to maximize its advantages.

As can be concluded, the use of various seaweed species (being green, red or brown) has the potential to enhance various qualities on poultry eggs. Such as quality, weight, yolk cholesterol reduction and, depending on the species, other bioactive molecules capable even of reducing toxic bacterium levels in the digestive system of poultries. It appears that a mixture between brown, green and red seaweed could be a promising supplement used in order to enrich eggs. However, such a product would need R&D work in order to determine the bioavailability of the molecules in a seaweed mixture. Once again, the concentration of the seaweed in the feed seems to be crucial.

3.5. Ruminat Feed

The use of seaweed in ruminant feeds has been affected by the high demand of animal feed protein, the need for alternatives to the traditional soybean and animal protein feed as well as the food market regulations related with the livestock feeding. Studies carried out to date regarding the use of seaweed in bovine, caprine and other ruminant nutrition have focused on the addition of small quantities of different macroalgal species to the feed and the subsequent assessment of the animal to check for possible prebiotic activity and enhanced animal performance.

Information on the application of green seaweeds in ruminant feed is scarce. *Ulva lactuca* could be fed to male lambs at up to 20% of diet, without negatively affecting the palatability. It presents low protein degradability (40%) and a moderate energy digestibility (60%), being comparable to a medium to low quality forage and suitable to use with feeds that have high energy/low protein content as cereal grains [119]. *Chaetomorpha linum* (Chlorophyta) was also used to feed growing lambs, with a 20% seaweed meal, having a slightly depressing effect on growth and feed conversion ratio, possibly due to the high ash content [9,120].

Red seaweed has received more attention, as demonstrated before, in bovine feed than in other ruminant feed [9]. There are some uses of red seaweed (a 70% concentrate of *Phymatolithon calcareum*—as *Lithothamnion calcareum*—extract fed at a ratio of 0.5 g/kg) with success in buffering the rumen pH, but they did not improve fiber digestion nor modify rumen fermentation [9,121]. This is in agreement with the literature, since the genus *Ulva* presents low ash levels (Table 1), which allows this seaweed to become a great option for future studies with bioavailability.

For example, supplementation of the brown seaweed *Ascophyllum nodosum* to feedlot cattle was found to reduce fecal shedding of *Esherichia coli* [2,121]. There is more research with the inclusion of seaweed in caprine feed. Orkney sheep, from the North Ronaldsay Island, are known to feed mostly brown seaweed most of the year. Species like *Laminaria digitata*, *Laminaria hyperborea* and *Saccharina latissima* (Phaeophyceaea) accounts for 90% of the summer feed of this sheep, meeting a substantial amount of nutrient requirements since they may have up to 13% crude protein. Orkney sheep also consume another seaweed species, like *Alaria esculenta*, *Ascophyllum nodosum*, Fucus sp. (brown seaweed), *Palmaria palmata* (red seaweed) and some green algae. Sheep consume seaweed in such quantity to sustain maintenance requirements but suffer from mineral overload due to its high mineral content [2,122]. There are also some studies suggesting the use of *Macrocystis pyrifera* up to 30% levels as a supplement in goat feed without affecting digestibility, degradability and parameters of ruminal fermentation (such as pH and ammoniac nitrogen). It was also noticeable the increase of rumen pH, water intake and urine excretion [2,122]. Species from the Genus Sargassum are also studied for this purpose. Nowadays, we know that it could be introduced at up to 30% in the diets of growing sheep and goats without depressing intake, growth performance and diet digestibility [59,123,124]. Eating *Sargassum* sp. increased water consumption, probably due to their high concentration in minerals, mainly Na and K, which could make *Sargassum* less suitable for feeding during dry periods. *Sargassum* sp. meal could be used to limit the decrease in rumen pH resulting from acidogenic diets. It also tended to decrease the concentration of volatile fatty acids [124]. Further research can incorporate the determination of bioavailability in a mixture between *Ulva lactuca* and one or more of the options mentioned in terms of red seaweed. This will allow us to understand if such a mixture can be used as a prebiotic, retaining the advantages of both species present in the mixture.

There are various observations of using mainly brown and red seaweed as ruminant feeds. However, the data is scarce with the exception of few punctual cases and is not enough to start R&D work in order to develop new products for the ruminant feed market. There are a lot of studies to be developed in order to sustain seaweed as feed supplement in ruminants.

Asparagopsis armata: The Future for Methane Emissions Reduction from Ruminant Animals?

The red seaweed *Asparagopsis armata* is one of the best hypothesis exploited to ameliorate one of the main problems that livestock farms face nowadays: high rates of enteric methane emissions [125]. Enteric methane is a natural by-product of microbial fermentation of nutrients in the digestive tract of animals [126]. There are considerable differences in contribution of enteric methane in different regions and countries of the world. For instance, it was estimated in 2017 that the enteric methane emissions from livestock, the main source of anthropogenic methane emissions in the US, reached 6.46 million tons, which is equivalent to 27% of the nation's anthropogenic emissions [127,128].

Seaweed has been a traditional part of the livestock diet and they have a historical usage in agriculture [1]. There have been several studies on seaweeds to characterize their effects as livestock feeds and their potential to manipulate rumen fermentation and methane production, which determined that the formulation of the basal feed is of key importance. Many seaweed species have been demonstrated to reduce methane production by rumen methanogens but with variable effects on fermentative health and substrate digestibility [129]. The *A. armata* is the only seaweed that demonstrated to remain effective and dramatically anti-methanogenic without negative impacts on rumen function and, at low inclusion levels, in animal diets [126,130,131]. Most of the initial breakthroughs in the inclusion of *A. armata* as livestock feed occurred in in vitro studies, all of which have demonstrated significant reduction of methane emissions at levels of approximately 2% of diet substrates [132–134]. Although it was considered that this dietary level of the seaweed was low and considered feasible for livestock production systems, in 2018 it was proved to be potentially effective at lower intake levels. Their study in sheep using *Asparagopsis taxiformis* reported up to 80% reduction of methane emission. This research was also important because of the observation of the refusal of the tested animals in the assay to ingest meals with high levels of seaweed, proving the potential of low intake levels [18]. The potential of the seaweed *Asparagopsis armata* to reduce methane emissions shown in in vitro studies was recently investigated in vivo using lactating dairy cows, thus evaluating the methane emission alongside the impact of seaweed on the quality of the milk produced. Adding *A. armata* at 0.5% reduced the methane production by 26.4%, by 20.5% in methane yield (adjusted for feed intake) and by 26.8% in methane intensity (adjusted for milk production) without compromising milk yield or intake. Increasing to a level of 1% resulted in reductions of 67.2% methane production, 42.6% methane yield and 60.0% methane intensity. However, feed intake and milk yield were also reduced. Bromoform concentration in milk was not significantly different in cows that consumed seaweed compared to control. Other mineral concentrations in milk may be increased so some processing may be necessary for *A. armata* to be used as a feed additive [126].

The work of Kinley et al. [135] demonstrated the effectiveness of *Asparagopsis* sp. in beef cattle fed with a high grain diet. The conclusions of this work point out *Asparagopsis* sp. included in the diet at 0.05, 0.10 and 0.20% and resulted in decrease of methane production (g/kg dry matter intake) of 9, 38 and 98%, respectively. Enteric H_2 emissions increased with increasing *Asparagopsis* inclusion by 0, 380 and 1700% without compromising feed intake. Growth rate of the steers was enhanced by the 0.10 and 0.20% inclusion levels after 90 days finishing period with average daily weight growth increases of 26 and 22%, respectively. Including *Asparagopsis* sp. in the concluding 60 days those values were enhanced by 51 and 42%, respectively.

This demonstrates that *Aspargopsis* sp. can be a player key for reducing the methane emission in the ruminants, without secondary effects; however, the seaweed needs to be added to the normal feed as feed supplement to be effective. The effect of *A. armata* is not only methane reduction; it can also supply important minerals for the ruminant growth and dietary digestibility, but the data regarding the last one is scarce, thus there is a need for more studies in this area. One of the main topics which should be studied is related with the bioavailability of metabolites and minerals during the livestock digestion, being that this topic is of high importance in order to determine exactly which one, metabolites and/or minerals, are responsible for the described effects.

3.6. Other Animals Feeds

There has been research with the purpose of including seaweeds in the diet of other species. It is also worth mentioning the research made with rabbit feed. Mainly, there are some encouraging results with the inclusion of red and green seaweed species in the feed.

Low amounts of green seaweed, mainly from the *Ulva* genus, also showed encouraging results. A meal with 1% *Ulva* showed positive effects on growth performance and diet digestibility, at the same time as no hematological or biochemical parameters show negative effects on rabbit health [136]. The inclusion of higher than 5% rates is usually associated with no statistical differences between control groups and seaweed feed groups [137–140]. The potential of seaweed as rabbit feed requires more study to assess its full potential. The usage of calcified red seaweed, such as *Lithothamnium* sp. up to 1% mix could lead to the increase of calcium in rabbits, inferred by the observation of a reduction in width and length of the intestinal villi [9,141–143]. In both cases, there is a tendency to use lower amounts of seaweed, since the benefits disappear as the percentage is increased. The results also show potential research in bioavailability on mixing the two species in order to retain the best advantages of both in one prebiotic solution.

It is worth mentioning the use of seaweed, mainly brown algae in pig farms. Historically, it is described as the usage of a mixture of brown algae species (boiled or raw), like *Fucus vesiculosus*, *Pelvetia* sp. or *A. nodosum* with cereal meal to fatten pigs in Sweden and Scotland [144,145]. However, it was already proved that high amounts of brown seaweed can be detrimental to pigs, such as causing weight loss after several weeks feeding them with a 10% *A. nodosum* [146]. This kind of result led us to using seaweed as an addictive in low amounts (1–2%) for potential benefits in pigs' health and meat quality [9]. There are two main reasons to use seaweed on pig feed. One of them is the use of seaweed as a prebiotic and its health effects. The use of seaweed and seaweed extracts have been shown to have prebiotic effects and enhance immunologic function in pigs and have been assessed to replace antibiotics in pig farms [9]. There is proof that the use of polysaccharides as fucoidan and laminarin as an extract improves piglet performance, being that laminarin is the main source for gut health and performance improvements [147,148]. On the other hand, a few studies have been done with raw seaweed. There was a Japanese team trial, in which they fed pigs with 0.8% unspecified seaweed species feed for four days, from 76 day- to 80 day-old subjects, resulting in Immunoglobulin A production in saliva and immune function [149]. The work of Dierick et al. [150,151] tried to reproduce their own in vitro results, which indicated that a 1% *A. nodosum* feed had a depressive effect on the gut flora, especially *E. coli*, while increasing the *Lactobacilli/E. coli* ratio and leading to resistance to intestinal disorders. However, seaweed meal added at 0.25, 0.5 and 1% to piglet diets, failed to enhance performance, gut health, plasma oxidative status and did not alter microbial ecology in the foregut and in the caecum [150,151]. It was latter theorized that this lack of effect may be due to phlorotannins in *A. nodosum*, which could counteract the prebiotic effects of other compounds due to a too low inclusion rate. Regarding the unchanged oxidative status, antioxidant vitamins in the diet may mask the antioxidant effect of seaweed [9,152]. Another main reason to use seaweed as pig feed is a proposed strategy to face the problem of lack of iodine in some populations. The usage of brown seaweed, such as *Laminaria* and *Ascophyllum*, to enrich pig's meat with organic iodine, which is readily metabolized and stored in pig muscle, is an easily controllable strategy, with no risk of overdosing, however limited, to achieve the referred propose [150,153]. Feed pigs with 2% of dried *A. nodosum* meal increased the concentration of iodine in the tissue by 2.7 to 6.8, depending on the tissue [150]. As said before, however limited, this strategy offers a solution by introducing iodine enriched food in human nutrition with a low risk.

There is some background of the use of seaweeds for diverse types of animals during human history, where the good results are mainly due to seaweed minerals and polysaccharides, enhancing the growth performance and potentiating digestibility of the normal diet, but the data are scarce, thus there is a need for more research in this area. One of the main topics which should be studied is related with the bioavailability of seaweed molecules during the livestock digestion. Such bioavailability studies are fundamental since it will allow not only to determine how much of the seaweed content is made available during the livestock digestion process but will also help to study the mechanisms in which some seaweed metabolites could counteract the effects of others in the same species or in a mixture of different seaweeds.

4. Conclusions

Seaweeds are close to becoming popular, due to their suitability as potential feedstock production, as well as supplements for food items. Seaweeds are rich in protein, dietary fibers and phytochemicals used to enhance the nutritional quality of animal feed. The increasing demand over renewable and sustainable energy sources without compromising on food and land resources can be fulfilled by seaweeds as they are fast growing, high biomass yielding with elevated and free of charge productivity, compared to other conventional biomass feedstock, such as corn or soybean.

The seaweed animal feed assays occur mainly as fresh seaweed, dried seaweed or even seaweed crude extract. There is a general lack of nutritional and biochemical studies of seaweed as feeds that makes the analysis of seaweed composition effect in the animal welfare difficult. Thus, more studies, regarding seaweed complete biochemical profile (macro and micronutrients, also seaweed metabolites), are needed to fully understand the impact of seaweeds in the animals.

However, seaweeds evidenced their potential to be further explored as an animal feed additive/supplement and cannot be applied as a complete substitute of the typical animal feed. Seaweed benefic effects are generally below 10% of the total concentration in the animal feed; above that, it was demonstrated to show negative effects and even animals refused to eat the provided feed.

Actually, with the active search for alternatives to the typical feed supplements and antibiotics, seaweed is one of the main hypotheses for animal feed supplementation, because seaweed production does not compete for arable land or fresh water. However, the wild seaweed biomass does not have a quality guarantee, because of the variations of nutritional values and risks of bioaccumulation of heavy metals, to provide a reliable source of safe animal feed supplementation. Consequently, seaweed aquaculture is the alternative solution for seaweed production and can be met through improvements in existing technology (already in use in Acadian Seaplants, seaweed Production Company from Canada).

Author Contributions: Conceived and designed the idea: T.M., A.I., T.C., M.M., J.C., K.B.; Organization of the team: J.C.; Writing and bibliographic research: T.M., A.I., T.C., M.M., J.C., K.B.; Supervision and Manuscript Revision, L.P. and K.B. All authors have read and agreed to the published version of the manuscript.

Funding: This work is financed by national funds through FCT—Foundation for Science and Technology, I.P., within the scope of the projects UIDB/04292/2020—MARE—Marine and Environmental Sciences Centre. João Cotas thanks the European Regional Development Fund through the Interreg Atlantic Area Program, under the project NASPA.

Conflicts of Interest: The authors declare no conflict of interest.

References

1. Evans, F.D.; Critchley, A.T. Seaweeds for animal production use. *J. Appl. Phycol.* **2014**, *26*, 891–899. [CrossRef]
2. Makkar, H.P.S.; Tran, G.; Heuzé, V.; Giger-Reverdin, S.; Lessire, M.; Lebas, F.; Ankers, P. Seaweeds for livestock diets: A review. *Anim. Feed Sci. Technol.* **2016**, *212*, 1–17. [CrossRef]
3. Jamal, P.; Olorunnisola, K.S.; Jaswir, I.; Tijani, I.D.R.; Ansari, A.H. Bioprocessing of seaweed into protein enriched feedstock: Process optimization and validation in reactor. *Int. Food Res. J.* **2017**, *24*, 382–386.
4. Corino, C.; Modina, S.C.; Di Giancamillo, A.; Chiapparini, S.; Rossi, R. Seaweeds in pig nutrition. *Animals* **2019**, *9*, 1126. [CrossRef] [PubMed]
5. Øverland, M.; Mydland, L.T.; Skrede, A. Marine macroalgae as sources of protein and bioactive compounds in feed for monogastric animals. *J. Sci. Food Agric.* **2019**, *99*, 13–24. [CrossRef] [PubMed]
6. Probst, L.; Frideres, L.; Pedersen, B.; Amato, F. *Sustainable, Safe and Nutritious Food: New Nutrient Sources*; Publications Office of the European Union: Luxembourg, 2015; p. 17.
7. Norziah, M.H.; Ching, C.Y. Nutritional composition of edible seaweed *Gracilaria changgi*. *Food Chem.* **2000**, *68*, 69–76. [CrossRef]
8. Mišurcová, L.; Machů, L.; Orsavová, J. Seaweed minerals as nutraceuticals. In *Advances in Food and Nutrition Research*; Elsevier: Amsterdam, The Netherlands, 2011; pp. 371–390.
9. Misurcova, L. Isolation and chemical properties of molecules derived from seaweeds chemical composition of seaweeds. *Handb. Mar. Macroalgae* **2011**, 171–192. [CrossRef]
10. Dawczynski, C.; Schubert, R.; Jahreis, G. Amino acids, fatty acids, and dietary fibre in edible seaweed products. *Food Chem.* **2007**, *103*, 891–899. [CrossRef]
11. Moreda-Piñeiro, J.; Moreda-Piñeiro, A.; Romarís-Hortas, V.; Domínguez-González, R.; Alonso-Rodríguez, E.; López-Mahía, P.; Muniategui-Lorenzo, S.; Prada-Rodríguez, D.; Bermejo-Barrera, P. Trace metals in marine foodstuff: Bioavailability estimation and effect of major food constituents. *Food Chem.* **2012**, *134*, 339–345. [CrossRef]
12. Gaillard, C.; Bhatti, H.S.; Novoa-Garrido, M.; Lind, V.; Roleda, M.Y.; Weisbjerg, M.R. Amino acid profiles of nine seaweed species and their in situ degradability in dairy cows. *Anim. Feed Sci. Technol.* **2018**, *241*, 210–222. [CrossRef]
13. Angell, A.R.; Angell, S.F.; de Nys, R.; Paul, N.A. Seaweed as a protein source for mono-gastric livestock. *Trends Food Sci. Technol.* **2016**, *54*, 74–84. [CrossRef]
14. Maehre, H.K.; Malde, M.K.; Eilertsen, K.-E.; Elvevoll, E.O. Characterization of protein, lipid and mineral contents in common Norwegian seaweeds and evaluation of their potential as food and feed. *J. Sci. Food Agric.* **2014**, *94*, 3281–3290. [CrossRef] [PubMed]
15. Maehre, H.K. Seaweed Proteins—How to Get to Them? Ph.D. Thesis, Arctic University of Norway, Tromsø, Norway, 2015.
16. Li, Q.; Luo, J.; Wang, C.; Tai, W.; Wang, H.; Zhang, X.; Liu, K.; Jia, Y.; Lyv, X.; Wang, L.; et al. Ulvan extracted from green seaweeds as new natural additives in diets for laying hens. *J. Appl. Phycol.* **2018**, *30*, 2017–2027. [CrossRef]
17. Azenha, I. Cultivo e Avaliação Nutricional de Ulva sp. Comparação com Exemplares Recolhidos em Ambiente Natural. Bachelor's Thesis, Escola Superior Agrária de Coimbra, Coimbra, Portugal, 2019.
18. Pereira, L. Biological and therapeutic properties of the seaweed polysaccharides. *Int. Biol. Rev.* **2018**, *2*, 1–50. [CrossRef]
19. Guiry, M.D.; Guiry, G.M. Algaebase: Listing the World's Algae. Available online: https://www.algaebase.org/ (accessed on 27 June 2020).

20. Kidgell, J.T.; Magnusson, M.; de Nys, R.; Glasson, C.R.K. Ulvan: A systematic review of extraction, composition and function. *Algal Res.* **2019**, *39*, 101422. [CrossRef]

21. O'Sullivan, L.; Murphy, B.; McLoughlin, P.; Duggan, P.; Lawlor, P.G.; Hughes, H.; Gardiner, G.E. Prebiotics from marine macroalgae for human and animal health applications. *Mar. Drugs* **2010**, *8*, 2038–2064. [CrossRef]

22. Usov, A.I. Polysaccharides of the red algae. In *Advances in Carbohydrate Chemistry and Biochemistry*; Horton, D., Ed.; Academic Press: New York, NY, USA, 2011; pp. 115–217.

23. Koizumi, J.; Takatani, N.; Kobayashi, N.; Mikami, K.; Miyashita, K.; Yamano, Y.; Wada, A.; Maoka, T.; Hosokawa, M. Carotenoid profiling of a red seaweed *Pyropia yezoensis*: Insights into biosynthetic pathways in the order Bangiales. *Mar. Drugs* **2018**, *16*, 426. [CrossRef] [PubMed]

24. Peng, J.; Yuan, J.-P.; Wu, C.-F.; Wang, J.-H. Fucoxanthin, a marine carotenoid present in brown seaweeds and diatoms: Metabolism and bioactivities relevant to human health. *Mar. Drugs* **2011**, *9*, 1806–1828. [CrossRef]

25. Peng, Y.; Hu, J.; Yang, B.; Lin, X.P.; Zhou, X.F.; Yang, X.W.; Liu, Y. Chemical composition of seaweeds. In *Seaweed Sustainability: Food and Non-Food Applications*; Elsevier Inc.: Amsterdam, The Netherlands, 2015; pp. 79–124, ISBN 9780124199583.

26. Haberecht, S.; Wilkinson, S.; Roberts, J.; Wu, S.; Swick, R. Unlocking the potential health and growth benefits of macroscopic algae for poultry. *Worlds. Poult. Sci. J.* **2017**, *74*, 5–20. [CrossRef]

27. Ponce, M.; Zuasti, E.; Anguís, V.; Fernández-Díaz, C. Effects of the sulfated polysaccharide ulvan from *Ulva ohnoi* on the modulation of the immune response in Senegalese sole (*Solea senegalensis*). *Fish Shellfish Immunol.* **2020**, *100*, 27–40. [CrossRef]

28. Gupta, S.; Abu-Ghannam, N. Recent developments in the application of seaweeds or seaweed extracts as a means for enhancing the safety and quality attributes of foods. *Innov. Food Sci. Emerg. Technol.* **2011**, *12*, 600–609. [CrossRef]

29. Hamed, I.; Özogul, F.; Özogul, Y.; Regenstein, J.M. Marine bioactive compounds and their health benefits: A Review. *Compr. Rev. Food Sci. Food Saf.* **2015**, *14*, 446–465. [CrossRef]

30. Valente, L.M.P.; Rema, P.; Ferraro, V.; Pintado, M.; Sousa-Pinto, I.; Cunha, L.M.; Oliveira, M.B.; Araújo, M. Iodine enrichment of rainbow trout flesh by dietary supplementation with the red seaweed *Gracilaria vermiculophylla*. *Aquaculture* **2015**, *446*, 132–139. [CrossRef]

31. Holdt, S.L.; Kraan, S. Bioactive compounds in seaweed: Functional food applications and legislation. *J. Appl. Phycol.* **2011**, *23*, 543–597. [CrossRef]

32. Cian, R.E.; Drago, S.R.; De Medina, F.S.; Martínez-Augustin, O. Proteins and carbohydrates from red seaweeds: Evidence for beneficial effects on gut function and microbiota. *Mar. Drugs* **2015**, *13*, 5358–5383. [CrossRef] [PubMed]

33. Černá, M. *Seaweed Proteins and Amino Acids as Nutraceuticals*, 1st ed.; Elsevier Inc.: Amsterdam, The Netherlands, 2011; Volume 64, ISBN 9780123876690.

34. Pereira, L. A review of the nutrient composition of selected edible seaweeds. In *Seaweed: Ecology, Nutrient Composition and Medicinal Uses*; Pomin, V.H., Ed.; Nova Science Publishers, Inc.: New York, NY, USA, 2011; pp. 15–47, ISBN 978-1-61470-878-0.

35. Kodner, R.B.; Pearson, A.; Summons, R.E.; Knoll, A.H. Sterols in red and green algae: Quantification, phylogeny, and relevance for the interpretation of geologic steranes. *Geobiology* **2008**, *6*, 411–420. [CrossRef]

36. Hayes, M. *Seaweeds: A Nutraceutical and Health Food*; Elsevier Inc.: Amsterdam, The Netherlands, 2015; ISBN 9780124199583.

37. Sánchez-Machado, D.I.; López-Cervantes, J.; López-Hernández, J.; Paseiro-Losada, P. Fatty acids, total lipid, protein and ash contents of processed edible seaweeds. *Food Chem.* **2004**, *85*, 439–444. [CrossRef]

38. Sá, A.R. Illustrated Guide of Macroalgae of the Bay of Buarcos. Master's Thesis, University of Coimbra, Coimbra, Portugal, 2019.

39. Pomin, V.H. *Seaweed: Ecology, Nutrient Composition, and Medicinal Uses*; Nova Science Publishers: New York, NY, USA, 2012; ISBN 9781614708780.

40. Laudadio, V.; Lorusso, V.; Lastella, N.M.B.; Dhama, K.; Karthik, K.; Tiwari, R.; Alam, G.M.; Tufarelli, V. Enhancement of Nutraceutical Value of Table Eggs Through Poultry Feeding Strategies. *Int. J. Pharmacol.* **2015**, *11*, 201–212. [CrossRef]

41. Vilà, B. *Improvement of Biologic and Nutritional Value of Eggs*; CIHEAM Options Méditerranéennes: Montpellier, France, 2008; Volume 390.

42. Cherry, P.; O'hara, C.; Magee, P.J.; Mcsorley, E.M.; Allsopp, P.J. Risks and benefits of consuming edible seaweeds. *Nutr. Rev.* **2019**, *77*, 307–329. [CrossRef]

43. Why Feed Safety?|Global Feed Safety Platform|Food and Agriculture Organization of the United Nations. Available online: http://www.fao.org/feed-safety/background/why-feed-safety/en/ (accessed on 18 July 2020).

44. Chen, Q.; Pan, X.D.; Huang, B.F.; Han, J.L. Distribution of metals and metalloids in dried seaweeds and health risk to population in southeastern China. *Sci. Rep.* **2018**, *8*, 1–7. [CrossRef]

45. Desideri, D.; Cantaluppi, C.; Ceccotto, F.; Meli, M.A.; Roselli, C.; Feduzi, L. Essential and toxic elements in seaweeds for human consumption. *J. Toxicol. Environ. Health Part A Curr. Issues* **2016**, *79*, 112–122. [CrossRef] [PubMed]

46. Bampidis, V.; Bastos, M.; Christensen, H.; Dusemund, B.; Kouba, M.; Kos Durjava, M.; López-Alonso, M.; López Puente, S.; Marcon, F.; Mayo, B.; et al. Guidance on the assessment of the safety of feed additives for the environment. *EFSA J.* **2019**, *17*. [CrossRef]

47. Lovell, R.T. Diet and fish husbandry. In *Fish Nutrition*; Elsevier: Amsterdam, The Netherlands, 2003; pp. 703–754.

48. Al-Asgah, N.A.; Younis, E.S.M.; Abdel-Warith, A.W.A.; Shamlol, F.S. Evaluation of red seaweed *Gracilaria arcuata* as dietary ingredient in African catfish, *Clarias gariepinus. Saudi J. Biol. Sci.* **2016**, *23*, 205–210. [CrossRef] [PubMed]

49. Glencross, B.D.; Booth, M.; Allan, G.L. A feed is only as good as its ingredients—A review of ingredient evaluation strategies for aquaculture feeds. *Aquac. Nutr.* **2007**, *13*, 17–34. [CrossRef]

50. Borquez, A.; Serrano, E.; Dantagnan, P.; Carrasco, J.; Hernandez, A. Feeding high inclusion of whole grain white lupin (*Lupinus albus*) to rainbow trout (*Oncorhynchus mykiss*): Effects on growth, nutrient digestibility, liver and intestine histology and muscle fatty acid composition. *Aquac. Res.* **2011**, *42*, 1067–1078. [CrossRef]

51. Francis, G.; Makkar, H.P.S.; Becker, K. Antinutritional factors present in plant-derived alternate fish feed ingredients and their effects in fish. *Aquaculture* **2001**, *199*, 197–227. [CrossRef]

52. Geurden, I.; Cuvier, A.; Gondouin, E.; Olsen, R.E.; Ruohonen, K.; Kaushik, S.; Boujard, T. Rainbow trout can discriminate between feeds with different oil sources. *Physiol. Behav.* **2005**, *85*, 107–114. [CrossRef]

53. Hardy, R.W. Alternate protein sources for salmon and trout diets. *Anim. Feed Sci. Technol.* **1996**, *59*, 71–80. [CrossRef]

54. Drew, M.D.; Borgeson, T.L.; Thiessen, D.L. A review of processing of feed ingredients to enhance diet digestibility in finfish. *Anim. Feed Sci. Technol.* **2007**, *138*, 118–136. [CrossRef]

55. Al-Hafedh, Y.S.; Alam, A.; Buschmann, A.H.; Fitzsimmons, K.M. Experiments on an integrated aquaculture system (seaweeds and marine fish) on the Red Sea coast of Saudi Arabia: Efficiency comparison of two local seaweed species for nutrient biofiltration and production. *Rev. Aquac.* **2012**, *4*, 21–31. [CrossRef]

56. FAO. *The State of World Fisheries and Aquaculture—Opportunities and Challenges*; FAO: Rome, Italy, 2014; ISBN 9789251082751.

57. Khalil, M.T.; El-Rakman, N.S.A. Abundance and diversity of surface zooplankton in the Gulf of Aqaba, Red Sea, Egypt. *J. Plankton Res.* **1997**, *19*, 927–936. [CrossRef]

58. Baars, M.; Schalk, P.; Veldhuis, J. easonal fluctuations in plankton biomass and productivity in the ecosystems of the Somali Current, Gulf of Aden, and southern Red Sea: Large marine ecosystems of the Indian Ocean: Assessment, sustainability, and management. In *Large Marine Ecosystems of the Indian Ocean: Assessment, Sustainability, and Management.*; Sherman, K., Okemwa, E.N., Ntiba, M.J., Eds.; Blackwell Science: Malden, MA, USA, 1998; pp. 143–174.

59. Pereira, R.; Yarish, C. Mass production of marine macroalgae. *Encycl. Ecol. Five-Volume Set* **2008**, 2236–2247. [CrossRef]

60. Casas-Valdez, M.; Hernández-Contreras, H.; Marín-Álvarez, A.; Aguila-Ramírez, R.N.; Hernández-Guerrero, C.J.; Sánchez-Rodríguez, I.; Carrillo-Domínguez, S. The seaweed *Sargassum* (Sargassaceae) as tropical alternative for goats' feeding. *Rev. Biol. Trop.* **2006**, *54*, 83–92. [CrossRef] [PubMed]

61. Nakagawa, H.; Umino, T.; Tasaka, Y. Usefulness of *Ascophyllum* meal as a feed additive for red sea bream, Pagrus major. *Aquaculture* **1997**, *151*, 275–281. [CrossRef]

62. Davies, S.J.; Brown, M.T.; Camilleri, M. Preliminary assessment of the seaweed *Porphyra purpurea* in artificial diets for thick-lipped grey mullet (*Chelon labrosus*). *Aquaculture* **1997**, *152*, 249–258. [CrossRef]

63. Wassef, E.A.; El Masry, M.H.; Mikhail, F.R. Growth enhancement and muscle structure of striped mullet, *Mugil cephalus* L., fingerlings by feeding algal meal-based diets. *Aquac. Res.* **2001**, *32*, 315–322. [CrossRef]

64. Ma, W.C.J.; Chung, H.Y.; Ang, P.O.; Kim, J.-S. Enhancement of bromophenol levels in aquacultured Silver Seabream (*Sparus sarba*). *J. Agric. Food Chem.* **2005**, *53*, 2133–2139. [CrossRef]

65. Valente, L.M.P.; Gouveia, A.; Rema, P.; Matos, J.; Gomes, E.F.; Pinto, I.S. Evaluation of three seaweeds *Gracilaria bursa-pastoris*, *Ulva rigida* and *Gracilaria cornea* as dietary ingredients in European sea bass (*Dicentrarchus labrax*) juveniles. *Aquaculture* **2006**, *252*, 85–91. [CrossRef]

66. Pham, M.A.; Lee, K.-J.; Lee, B.-J.; Lim, S.-J.; Kim, S.-S.; Lee, Y.-D.; Heo, M.-S.; Lee, K.-W. Effects of dietary *Hizikia fusiformis* on growth and immune responses in juvenile Olive Flounder (*Paralichthys olivaceus*). *Asian-Australas. J. Anim. Sci.* **2006**, *19*, 1769–1775. [CrossRef]

67. Kalla, A.; Yoshimatsu, T.; Araki, T.; Zhang, D.-M.; Yamamoto, T.; Sakamoto, S. Use of *Porphyra* spheroplasts as feed additive for red sea bream. *Fish. Sci.* **2008**, *74*, 104–108. [CrossRef]

68. Khan, M.N.D.; Yoshimatsu, T.; Kalla, A.; Araki, T.; Sakamoto, S. Supplemental effect of *Porphyra* spheroplasts on the growth and feed utilization of black sea bream. *Fish. Sci.* **2008**, *74*, 397–404. [CrossRef]

69. Soler-Vila, A.; Coughlan, S.; Guiry, M.D.; Kraan, S. The red alga *Porphyra dioica* as a fish-feed ingredient for rainbow trout (*Oncorhynchus mykiss*): Effects on growth, feed efficiency, and carcass composition. *J. Appl. Phycol.* **2009**, *21*, 617–624. [CrossRef]

70. Ergün, S.; Soyutürk, M.; Güroy, B.; Güroy, D.; Merrifield, D. Influence of *Ulva* meal on growth, feed utilization, and body composition of juvenile Nile tilapia (*Oreochromis niloticus*) at two levels of dietary lipid. *Aquac. Int.* **2009**, *17*, 355–361. [CrossRef]

71. Güroy, D.; Güroy, B.; Merrifield, D.L.; Ergün, S.; Tekinay, A.A.; Yiğit, M. Effect of dietary *Ulva* and *Spirulina* on weight loss and body composition of rainbow trout, *Oncorhynchus mykiss* (Walbaum), during a starvation period. *J. Anim. Physiol. Anim. Nutr. (Berl).* **2011**, *95*, 320–327. [CrossRef]

72. Güroy, B.; Ergün, S.; Merrifield, D.L.; Güroy, D. Effect of autoclaved *Ulva* meal on growth performance, nutrient utilization and fatty acid profile of rainbow trout, Oncorhynchus mykiss. *Aquac. Int.* **2013**, *21*, 605–615. [CrossRef]

73. Kamunde, C.; Sappal, R.; Melegy, T.M. Brown seaweed (AquaArom) supplementation increases food intake and improves growth, antioxidant status and resistance to temperature stress in Atlantic salmon, *Salmo salar*. *PLoS ONE* **2019**, *14*, 1–24. [CrossRef]

74. Ferreira, M.; Larsen, B.K.; Granby, K.; Cunha, S.C.; Monteiro, C.; Fernandes, J.O.; Nunes, M.L.; Marques, A.; Dias, J.; Cunha, I.; et al. Diets supplemented with *Saccharina latissima* influence the expression of genes related to lipid metabolism and oxidative stress modulating rainbow trout (*Oncorhynchus mykiss*) fillet composition. *Food Chem. Toxicol.* **2020**, *140*, 111332. [CrossRef]

75. Utting, S.D.; Millican, P.F. Techniques for the hatchery conditioning of bivalve broodstocks and the subsequent effect on egg quality and larval viability. *Aquaculture* **1997**, *155*, 45–54. [CrossRef]

76. Gonzalez Araya, R.; Mingant, C.; Petton, B.; Robert, R. Influence of diet assemblage on *Ostrea edulis* broodstock conditioning and subsequent larval development. *Aquaculture* **2012**, *364–365*, 272–280. [CrossRef]

77. Gallager, S.M.; Mann, R. Growth and survival of larvae of *Mercenaria mercenaria* (L.) and *Crassostrea virginica* (Gmelin) relative to broodstock conditioning and lipid content of eggs. *Aquaculture* **1986**, *56*, 105–121. [CrossRef]

78. Pronker, A.E.; Nevejan, N.M.; Peene, F.; Geijsen, P.; Sorgeloos, P. Hatchery broodstock conditioning of the blue mussel *Mytilus edulis* (Linnaeus 1758). Part I. Impact of different micro-algae mixtures on broodstock performance. *Aquac. Int.* **2008**, *16*, 297–307. [CrossRef]

79. González-Araya, R.; Quéau, I.; Quéré, C.; Moal, J.; Robert, R. A physiological and biochemical approach to selecting the ideal diet for *Ostrea edulis* (L.) broodstock conditioning (part A). *Aquac. Res.* **2011**, *42*, 710–726. [CrossRef]

80. González-Araya, R.; Lebrun, L.; Quéré, C.; Robert, R. The selection of an ideal diet for *Ostrea edulis* (L.) broodstock conditioning (part B). *Aquaculture* **2012**, *362–363*, 55–66. [CrossRef]

81. Anjos, C.; Baptista, T.; Joaquim, S.; Mendes, S.; Matias, A.M.; Moura, P.; Simões, T.; Matias, D. Broodstock conditioning of the Portuguese oyster (*Crassostrea angulata*, Lamarck, 1819): Influence of different diets. *Aquac. Res.* **2017**, *48*, 3859–3878. [CrossRef]

82. Brown, M.; Robert, R. Preparation and assessment of microalgal concentrates as feeds for larval and juvenile Pacific oyster (*Crassostrea gigas*). *Aquaculture* **2002**, *207*, 289–309. [CrossRef]

83. Coutteau, P.; Sorgeloos, P. The use of algal substitutes and the requirement for live algae in the hatchery and nursery rearing of bivalve molluscs: An international survey. *J. Shellfish Res.* **1992**, *11*, 118–128.

84. Boeing, P.; Escondido, C. Use of spray-dried *Schizochytrium* sp. as a partial algal replacement for juvenile bivalves. *J. Shellfish Res.* **1997**, *16*, 284.

85. Arney, B.; Liu, W.; Forster, I.P.; McKinley, R.S.; Pearce, C.M. Feasibility of dietary substitution of live microalgae with spray-dried *Schizochytrium* sp. or *Spirulina* in the hatchery culture of juveniles of the Pacific geoduck clam (*Panopea generosa*). *Aquaculture* **2015**, *444*, 117–133. [CrossRef]

86. Matias, D.; Joaquim, S.; Leitão, A.; Massapina, C. Effect of geographic origin, temperature and timing of broodstock collection on conditioning, spawning success and larval viability of *Ruditapes decussatus* (Linné, 1758). *Aquac. Int.* **2009**, *17*, 257–271. [CrossRef]

87. Joaquim, S.; Matias, D.; Matias, A.M.; Moura, P.; Arnold, W.S.; Chícharo, L.; Baptista Gaspar, M. Reproductive activity and biochemical composition of the pullet carpet shell *Venerupis senegalensis* (Gmelin, 1791) (Mollusca: Bivalvia) from Ria de Aveiro (northwestern coast of Portugal). *Sci. Mar.* **2010**, *75*, 217–226. [CrossRef]

88. Noël, L.; Chekri, R.; Millour, S.; Vastel, C.; Kadar, A.; Sirot, V.; Leblanc, J.C.; Guérin, T. Li, Cr, Mn, Co, Ni, Cu, Zn, Se and Mo levels in foodstuffs from the Second French TDS. *Food Chem.* **2012**, *132*, 1502–1513. [CrossRef]

89. Fleurence, J.; Morançais, M.; Dumay, J.; Decottignies, P.; Turpin, V.; Munier, M.; Garcia-Bueno, N.; Jaouen, P. What are the prospects for using seaweed in human nutrition and for marine animals raised through aquaculture? *Trends Food Sci. Technol.* **2012**, *27*, 57–61. [CrossRef]

90. Cardoso, C.; Afonso, C.; Lourenço, H.; Costa, S.; Nunes, M.L. Bioaccessibility assessment methodologies and their consequences for the risk-benefit evaluation of food. *Trends Food Sci. Technol.* **2015**, *41*, 5–23. [CrossRef]

91. Cardoso, C.; Gomes, R.; Rato, A.; Joaquim, S.; Machado, J.; Gonçalves, J.F.; Vaz-Pires, P.; Magnoni, L.; Matias, D.; Coelho, I.; et al. Elemental composition and bioaccessibility of farmed oysters (*Crassostrea gigas*) fed different ratios of dietary seaweed and microalgae during broodstock conditioning. *Food Sci. Nutr.* **2019**, *7*, 2495–2504. [CrossRef] [PubMed]

92. Zahid, P.B.; Ali, A.; Zahid, M.-J. Brown seaweeds as supplement for broiler feed. *Hamdard Med.* **2001**, *2*, 98–101.

93. Ali, A.; Memon, M.S. Green seaweed as component of poultry feed. *Int. J. Biol. Biotechn.* **2014**, *5*, 211–214.

94. Abudabos, A.M.; Okab, A.B.; Aljumaah, R.S.; Samara, E.M.; Abdoun, K.A.; Al-Haidary, A.A. Nutritional value of green seaweed (*Ulva Lactuca*) for broiler chickens. *Ital. J. Anim. Sci.* **2013**, *12*, e28. [CrossRef]

95. Wang, S.; Shi, X.; Zhou, C.; Lin, Y. *Entermorpha prolifera*: Effects on performance, carcass quality and small intestinal digestive enzyme activities of broilers. *Chin. J. Anim. Nutr.* **2013**, *25*, 1332–1337.

96. Wang, S.; Jia, Y.; Wang, L.; Zhu, F.; Lin, Y. *Enteromorpha prolifera* supplemental level: Effects on laying performance, egg quality, immune function and microflora in feces of laying hens. *Chin. J. Anim. Nutr.* **2013**, *25*, 1346–1352.

97. Asar, M. *The Use of Some Weeds in Poultry Nutrition*; University of Alexandria: Alexandria, Egypt, 1972.

98. El-Deek, A.A.; Al-Harthi, M.A.; Abdalla, A.A.; Elbanoby, M.M. The use of brown algae meal in finisher broiler diets. *Egypt. Poult. Sci* **2011**, *31*, 767–781.

99. Gu, H.Y.; Shu, Z.Z.; Liu, Y.G. Nutrient composition of marine algae and their feeding on broilers. *Chin. J. Anim. Sci.* **1988**, *3*, 12–14.

100. Ventura, M.R.; Castañon, J.I.R.; McNab, J.M. Nutritional value of seaweed (*Ulva rigida*) for poultry. *Anim. Feed Sci. Technol.* **1994**, *49*, 87–92. [CrossRef]

101. Zahid, P.B.; Aisha, K.; Ali, A. Green seaweeds as component of poultry feed. *Bangladesh J. Bot.* **1995**, *24*, 146–153.

102. Kulshreshtha, G.; Rathgeber, B.; Stratton, G.; Thomas, N.; Evans, F.; Critchley, A.; Hafting, J.; Prithiviraj, B. Immunology, health, and disease: Feed supplementation with red seaweeds, *Chondrus crispus* and *Sarcodiotheca gaudichaudii*, affects performance, egg quality, and gut microbiota of layer hens. *Poult. Sci.* **2014**, *93*, 2991–3001. [CrossRef] [PubMed]

103. Baurhoo, B.; Phillip, L.; Ruiz-Feria, C.A.A. Effects of purified lignin and mannan oligosaccharides on intestinal integrity and microbial populations in the ceca and litter of broiler chickens. *Poult. Sci.* **2007**, *86*, 1070–1078. [CrossRef] [PubMed]

104. Li, X.; Liu, L.; Li, K.; Hao, K.; Xu, C. Effect of fructooligosaccharides and antibiotics on laying performance of chickens and cholesterol content of egg yolk. *Br. Poult. Sci.* **2007**, *48*, 185–189. [CrossRef] [PubMed]

105. Awad, W.A.; Ghareeb, K.; Abdel-Raheem, S.; Böhm, J. Effects of dietary inclusion of probiotic and synbiotic on growth performance, organ weights, and intestinal histomorphology of broiler chickens. *Poult. Sci.* **2009**, *88*, 49–55. [CrossRef]

106. Filazi, A.; Sireli, U.T.; Cadirci, O. Residues of gentamicin in eggs following medication of laying hens. *Br. Poult. Sci.* **2005**, *46*, 580–583. [CrossRef]

107. Shams Shargh, M.; Dastar, B.; Zerehdaran, S.; Khomeiri, M.; Moradi, A. Effects of using plant extracts and a probiotic on performance, intestinal morphology, and microflora population in broilers. *J. Appl. Poult. Res.* **2012**, *21*, 201–208. [CrossRef]

108. Tellez, G.; Pixley, C.; Wolfenden, R.E.; Layton, S.L.; Hargis, B.M. Probiotics/direct fed microbials for *Salmonella* control in poultry. *Food Res. Int.* **2012**, *45*, 628–633. [CrossRef]

109. Kulshreshtha, G.; Rathgeber, B.; MacIsaac, J.; Boulianne, M.; Brigitte, L.; Stratton, G.; Thomas, N.A.; Critchley, A.T.; Hafting, J.; Prithiviraj, B. Feed supplementation with red seaweeds, *Chondrus crispus* and *Sarcodiotheca gaudichaudii*, reduce Salmonella Enteritidis in laying hens. *Front. Microbiol.* **2017**, *8*, 1–12. [CrossRef] [PubMed]

110. Wijesekara, I.; Pangestuti, R.; Kim, S.K. Biological activities and potential health benefits of sulfated polysaccharides derived from marine algae. *Carbohydr. Polym.* **2011**, *84*, 14–21. [CrossRef]

111. Ngo, D.H.; Kim, S.K. Sulfated polysaccharides as bioactive agents from marine algae. *Int. J. Biol. Macromol.* **2013**, *62*, 70–75. [CrossRef] [PubMed]

112. Fredriksson, S.; Elwinger, K.; Pickova, J. Fatty acid and carotenoid composition of egg yolk as an effect of microalgae addition to feed formula for laying hens. *Food Chem.* **2006**, *99*, 530–537. [CrossRef]

113. Anton, M. Composition and structure of hen egg yolk. *Bioact. Egg Compd.* **2007**, 1–6. [CrossRef]

114. Laudadio, V.; Ceci, E.; Lastella, N.M.B.; Introna, M.; Tufarelli, V. Low-fiber alfalfa (*Medicago sativa* L.) meal in the laying hen diet: Effects on productive traits and egg quality. *Poult. Sci.* **2014**, *93*, 1868–1874. [CrossRef]

115. Schiavone, A.; Barroeta, A.C. Egg enrichment with vitamins and trace minerals. In *Improving the Safety and Quality of Eggs and Egg Products*; Elsevier: Amsterdam, The Netherlands, 2011; pp. 289–320.

116. Ehr, I.J.; Persia, M.E.; Bobeck, E.A. Comparative omega-3 fatty acid enrichment of egg yolks from first-cycle laying hens fed flaxseed oil or ground flaxseed. *Poult. Sci.* **2017**, *96*, 1791–1799. [CrossRef]

117. Choi, Y.; Lee, E.C.; Na, Y.; Lee, S.R. Effects of dietary supplementation with fermented and non-fermented brown algae by-products on laying performance, egg quality, and blood profile in laying hens. *Asian-Australasian J. Anim. Sci.* **2018**, *31*, 1654–1659. [CrossRef]

118. Al-Harthi, M.A.; El-Deek, A.A. Effect of different dietary concentrations of brown marine algae (*Sargassum dentifebium*) prepared by different methods on plasma and yolk lipid profiles, yolk total carotene and lutein plus zeaxanthin of laying hens. *Ital. J. Anim. Sci.* **2012**, *11*, 347–353. [CrossRef]

119. Arieli, A.; Sklan, D.; Kissil, G. A note on the nutritive value of *Ulva lactuca* for ruminants. *Anim. Sci.* **1993**, *57*, 329–331. [CrossRef]

120. Rjiba Ktita, S.; Chermiti, A.; Mahouachi, M. The use of seaweeds (*Ruppia maritima* and *Chaetomorpha linum*) for lamb fattening during drought periods. *Small Rumin. Res.* **2010**, *91*, 116–119. [CrossRef]

121. Montañez-Valdez, O.D.; Pinos-Rodríguez, J.M.; Rojo-Rubio, R.; Salinas-Chavira, J.; Martíneztinajero, J.J.; Salem, A.Z.M.; Avellaneda-Cevallos, J.H. Effect of a calcified-seaweed extract as rumen buffer on ruminal disappearance and fermentation in steers. *Indian J. Anim. Sci.* **2012**, *82*, 430–432.

122. Hansen, H.; Hector, B.; Feldmann, J. A qualitative and quantitative evaluation of the seaweed diet of North Ronaldsay sheep. *Anim. Feed Sci. Technol.* **2003**, *105*, 21–28. [CrossRef]

123. Marín, A.; Casas, M.; Carrillo, S.; Hernández, H.; Monroy, A. Performance of sheep fed rations with *Sargassum* spp. sea algae. *Cuba. J. Agric. Sci.* **2003**, *37*, 119–123.

124. Marín, A.; Casas-Valdez, M.; Carrillo, S.; Hernández, H.; Monroy, A.; Sanginés, L.; Pérez-Gil, F. The marine algae *Sargassum* spp. (Sargassaceae) as feed for sheep in tropical and subtropical regions. *Rev. Biol. Trop.* **2009**, *57*, 1271–1281. [CrossRef]

125. Gerber, P.J.; Steinfeld, H.; Henderson, B.; Mottet, A.; Opio, C.; Dijkman, J.; Falcucci, A.; Tempio, G. A global Assessment of emissions and mitigation opportunities. In *Tackling Climate Change Through Livestock: A Global Assessment of Emissions and Mitigation Opportunities*; Food and Agriculture Organization of the United Nations: Rome, Italy, 2013; ISBN 9789251079201.

126. Roque, B.M.; Salwen, J.K.; Kinley, R.; Kebreab, E. Inclusion of *Asparagopsis armata* in lactating dairy cows' diet reduces enteric methane emission by over 50 percent. *J. Clean. Prod.* **2019**, *234*, 132–138. [CrossRef]

127. NASEM. *Improving Characterization of Anthropogenic Methane Emissions in the United States*; National Academies Press: Washington, DC, USA, 2018; ISBN 978-0-309-47050-6.

128. US EPA. Overview of Greenhouse Gases|Greenhouse Gas (GHG) Emissions|US EPA. Available online: https://www.epa.gov/ghgemissions/overview-greenhouse-gases (accessed on 16 June 2020).

129. Machado, L.; Magnusson, M.; Paul, N.A.; De Nys, R.; Tomkins, N. Effects of marine and freshwater macroalgae on in vitro total gas and methane production. *PLoS ONE* **2014**, *9*. [CrossRef]

130. Kinley, R.D.; de Nys, R.; Vucko, M.J.; Machado, L.; Tomkins, N.W. The red macroalgae *Asparagopsis taxiformis* is a potent natural antimethanogenic that reduces methane production during in vitro fermentation with rumen fluid. *Anim. Prod. Sci.* **2016**, *56*, 282. [CrossRef]

131. Li, X.; Norman, H.C.; Kinley, R.D.; Laurence, M.; Wilmot, M.; Bender, H.; de Nys, R.; Tomkins, N. *Asparagopsis taxiformis* decreases enteric methane production from sheep. *Anim. Prod. Sci.* **2018**, *58*, 681. [CrossRef]

132. Dubois, B.; Tomkins, N.W.D.; Kinley, R.; Bai, M.; Seymour, S.; Paul, N.A.; de Nys, R. Effect of tropical algae as additives on rumen in Vitro gas production and fermentation characteristics. *Am. J. Plant Sci.* **2013**, *4*, 34–43. [CrossRef]

133. Machado, L.; Magnusson, M.; Paul, N.A.; Kinley, R.; de Nys, R.; Tomkins, N. Dose-response effects of *Asparagopsis taxiformis* and *Oedogonium* sp. on in vitro fermentation and methane production. *J. Appl. Phycol.* **2016**, *28*, 1443–1452. [CrossRef]

134. Machado, L.; Magnusson, M.; Paul, N.A.; Kinley, R.; de Nys, R.; Tomkins, N. Identification of bioactives from the red seaweed *Asparagopsis taxiformis* that promote antimethanogenic activity in vitro. *J. Appl. Phycol.* **2016**, *28*, 3117–3126. [CrossRef]

135. Kinley, R.D.; Martinez-Fernandez, G.; Matthews, M.K.; de Nys, R.; Magnusson, M.; Tomkins, N.W. Mitigating the carbon footprint and improving productivity of ruminant livestock agriculture using a red seaweed. *J. Clean. Prod.* **2020**, *259*, 120836. [CrossRef]

136. El-banna, S.G.; Hassan, A.A.; Okab, A.B.; Koriem, A.A.; Ayoub, M.A. Effect of feeding diets supplemented with seaweed on growth performance and some blood hematological and biochemical characteristics of male Baladi rabbits. In Proceedings of the 4th International Conference on Rabbit Production in Hot Climates, Sharm Elsheikh, Egypt, 24–27 February 2005; pp. 373–382.

137. Raju, K.V.S.; Sreemannarayana, O. Feeding of *Ulva fasciata* to rabbits—Feed efficiency and carcass characteristics. *Indian Vet. J.* **1995**, *72*, 1331–1332.

138. Sreemannarayana, O.; Raju, K.V.S.; Ramaraju, G.V.A.N.S.; Prasad, J.R. The use of *Ulva fasciata*, a marine alga as rabbit feed: Growth and conversion efficiency. *Indian Vet. J.* **1995**, *72*, 989–991.

139. Chermiti, A.; Rjiba, S.; Mahouachi, M. Marine plants: A new alternative feed resource for livestock. In Proceedings of the 2009 FAO/IAEA International Symposium on Sustainable Improvement of Animal Production and Health, Vienna, Austria, 8–11 June 2009; p. 240.

140. Okab, A.B.; Samara, E.M.; Abdoun, K.A.; Rafay, J.; Ondruska, L.; Parkanyi, V.; Pivko, J.; Ayoub, M.A.; Al-Haidary, A.A.; Aljumaah, R.S.; et al. Effects of dietary seaweed (*Ulva lactuca*) supplementation on the reproductive performance of buck and doe rabbits. *J. Appl. Anim. Res.* **2013**, *41*, 347–355. [CrossRef]

141. Euler, A.C.C.; Ferreira, W.M.; Maurício, R.; Sousa, L.; Carvalho, W.; Teixeira, E.D.A.; Coelho, C.C.G.M.; Matos, C. In Vitro gas production of diets with inclusion of seaweed (*Lithothamnium* sp.) flour for white New Zealand Rabbits. In Proceedings of the 9th World Rabbit Congress, Verona, Italy, 10–13 June 2008; pp. 655–660.

142. Leroyer, J.; Coulombel, A. Chez Pascal et Myriam Orain, un atelier cunicole biologique qui fonctionne! *Cunicult. Mag.* **2009**, *36*, 1–4.

143. Euler, A.C.C.; Ferreira, W.M.; Teixeira, E.D.A.; Lana, A.; Guedes, R.M.C.; Avelar, A.C. Performance, digestibility and morphometry of ileal villi of rabbits fed with levels of inclusion of *Lithothamnium*. *Rev. Bras. Saúde e Produção Anim.* **2010**, *11*, 91–103.

144. Sauvageau, C. *Utilisation des Algues Marines*; Librairie Octave Doin: Paris, France, 1920.

145. Chapman, V.J.; Chapman, D.J. *Seaweeds and Their Uses*; Springer: Berlin/Heidelberg, Germany, 1980; ISBN 9400958064.

146. Jones, R.T.; Blunden, G.; Probert, A.J. Effects of dietary *Ascophyllum nodosum* on blood parameters of rats and pigs. *Bot. Mar.* **1979**, *22*, 393–402. [CrossRef]

147. Gahan, D.A.; Lynch, M.B.; Callan, J.J.; O'Sullivan, J.T.; O'Doherty, J.V. Performance of weanling piglets offered low-, medium- or high-lactose diets supplemented with a seaweed extract from *Laminaria* spp. *Animal* **2009**, *3*, 24–31. [CrossRef]

148. McDonnell, P.; Figat, S.; O'Doherty, J.V. The effect of dietary laminarin and fucoidan in the diet of the weanling piglet on performance, selected faecal microbial populations and volatile fatty acid concentrations. *Animal* **2010**, *4*, 579–585. [CrossRef] [PubMed]

149. Katayama, M.; Fukuda, T.; Okamura, T.; Suzuki, E.; Tamura, K.; Shimizu, Y.; Suda, Y.; Suzuki, K. Effect of dietary addition of seaweed and licorice on the immune performance of pigs. *Anim. Sci. J.* **2011**, *82*, 274–281. [CrossRef]

150. Dierick, N.; Ovyn, A.; De Smet, S. Effect of feeding intact brown seaweed *Ascophyllum nodosum* on some digestive parameters and on iodine content in edible tissues in pigs. *J. Sci. Food Agric.* **2009**, *89*, 584–594. [CrossRef]

151. Dierick, N.; Ovyn, A.; De Smet, S. In vitro assessment of the effect of intact marine brown macro-algae *Ascophyllum nodosum* on the gut flora of piglets. *Livest. Sci.* **2010**, *133*, 154–156. [CrossRef]

152. Michiels, J.; Skrivanova, E.; Missotten, J.; Ovyn, A.; Mrazek, J.; De Smet, S.; Dierick, N. Intact brown seaweed (*Ascophyllum nodosum*) in diets of weaned piglets: Effects on performance, gut bacteria and morphology and plasma oxidative status. *J. Anim. Physiol. Anim. Nutr.* **2012**, *96*, 1101–1111. [CrossRef] [PubMed]

153. Banoch, T.; Fajt, Z.; Drabek, J.; Svoboda, M. Iodine and its importance in human and pigs. *Veterinarstvi* **2010**, *60*, 690–694.

Journal of
*Marine Science
and Engineering*

Review

A Review of the Varied Uses of Macroalgae as Dietary Supplements in Selected Poultry with Special Reference to Laying Hen and Broiler Chickens

Garima Kulshreshtha [1,*], Maxwell T. Hincke [1,2], Balakrishnan Prithiviraj [3] and Alan Critchley [4]

[1]   Department of Cellular and Molecular Medicine, Faculty of Medicine, University of Ottawa, Ottawa, ON K1H 8M5, Canada; mhincke@uottawa.ca

[2]   Department of Innovation in Medical Education, Faculty of Medicine, University of Ottawa, Ottawa, ON K1H 8M5, Canada

[3]   Department of Plant, Food, and Environmental Sciences, Agricultural Campus, Dalhousie University, PO Box 550, Truro, NS B2N 5E3, Canada; bprithiviraj@dal.ca

[4]   Verschuren Centre for Sustainability in Energy and Environment, Cape Breton University, Sydney, Cape Breton, NS B1P 6L2, Canada; alan.critchley2016@gmail.com

*   Correspondence: gkulshre@uottawa.ca

Received: 27 June 2020; Accepted: 15 July 2020; Published: 19 July 2020

Abstract: Seaweeds comprise ca. 12,000 species. Global annual harvest is ca. 30.13 million metric tonnes, (valued ca. $11.7 billion USD in 2016) for various commercial applications. The growing scope of seaweed-based applications in food, agricultural fertilizers, animal feed additives, pharmaceuticals, cosmetics and personal care is expected to boost market demand. Agriculture and animal feed applications held the second largest seaweed market share in 2017, and the combined market is anticipated to reach much higher values by 2024 due to the impacts of current research and development targeting enhanced animal health and productivity. In general, seaweeds have been utilized in animal feed as a rich source of carbohydrates, protein, minerals, vitamins and dietary fibers with relatively well-balanced amino acid profiles and a unique blend of bioactive compounds. Worldwide, the animal nutrition market is largely driven by rising demand for poultry feeds, which represents ca. 47% of the total consumption for all animal nutrition. This review provides an overview of the utilization of specific seaweeds as sustainable feed sources for poultry production, including a detailed survey of seaweed-supplemented diets on growth, performance, gastrointestinal flora, disease, immunity and overall health of laying/broiler hens. Anti-microbial effects of seaweeds are also discussed.

Keywords: seaweed-supplemented feed; poultry; prebiotics; anti-microbial; gastrointestinal flora; immunity; animal nutrition market

1. Introduction

Algae comprise around 25,000–50,000 species, with a diversity of size, forms, pigments and functional compounds; ca. 12,000 of these are designated as macroalgae or seaweeds [1]. The global annual harvest of macroalgae is almost 36 million metric tonnes, with a market size of approximately $6 billion USD for various commercial applications. Global seaweed production is mainly carried out in Asian countries, which accounts for over 99% of global production [2]. Seaweeds can have high crop productivities per unit area as they do not require land and fresh water for growth, with lucrative scope for commercialization [3]. Seaweeds are a source of unique bioactive metabolites, which are not synthesized by terrestrial plants [4]. Bioactive molecules such as carbohydrates, proteins, minerals, polyphenols, pigments (chlorophylls, fucoxanthins, phycobilins), mycosporine-like amino acids (MAAs) and polyunsaturated fatty acids (PUFAs), including omega-3 fatty acids, have been attributed to various biological functionalities, such as anti-microbial, anti-viral, anti-inflammatory, immunomodulatory, prebiotic and cholesterol lowering effects [5]. Globally, seaweed cultivation has been growing rapidly, and it is currently produced in over 50 countries. According to the Food and Agriculture Organization (FAO), approximately 30.13 million tonnes of seaweeds were harvested in 2016 for various applications including direct consumption, food production, hydrocolloids, fertilizers and animal feed [6]. While seaweed bioactives are an appealing source for commercialization due to their various high value applications, the utilization of this resource has not been completely optimized.

Seaweeds have been utilized in animal feed as a rich source of carbohydrates, protein, minerals, vitamins and dietary fibers, with relatively well-balanced amino acid profiles and a unique blend of bioactive compounds. Recent developments in feed processing technologies have improved the nutritional quality of animal feed products [2,7]. The global market for animal feed additives and nutritional supplements was valued at 54 billion USD in 2018 and is estimated to generate a net revenue of 64 billion USD by 2025, growing at a compound annual growth rate (CAGR) of 2.7%. Worldwide, the animal nutrition market is largely driven by a rising demand for poultry feed, which constitutes about 47% of the total consumption [7].

There has been increasing interest in the market potential for functional feeds for livestock, with added-value linked to the health benefits for farm animals. Increasing consumer awareness regarding poultry meat quality, recent outbreaks associated with poultry diseases and the utilization of poultry meat and egg products as economical sources of protein are the major driving forces amplifying the animal feed additive market [8]. Worldwide, several seaweed companies, such as Aurora (Edmonds, WA), MBD (Melbourne, Australia), Alltech (Nicholasville, KY), Cellana (Kailua-Kona, HI), Ocean Harvest (County Galway, Ireland), Olmix (Bréhan, France), AquAgri (New Delhi, India) and ASL (NS, Canada), have been commercially producing high value seaweed-based commercial feed products for animal nutrition. These commercial products can potentially improve the health and performance of livestock animals with reduced investments in feed.

This review provides an overview of the utilization of various specific seaweeds as sustainable feed sources for poultry production. A detailed survey of seaweed-supplemented diets on growth, performance, gastrointestinal flora, disease, immunity and overall health of laying/broiler hens is presented. Conclusions drawn and potential future developments are also discussed, with the expectation that this review may open new opportunities to investigate enhanced exploitation for the potential of various, efficacious seaweeds, especially for sustainable growth in the poultry feed industry.

2. Effects of Various Seaweeds on Poultry Production

Collectively, the poultry industry has explored novel candidates of seaweeds as dietary supplements. A major goal of introducing supplements into the poultry diet is to enhance the efficacy of feed for the cost-effective production of commercially important meat and eggs, whilst also maintaining and/or improving poultry health.

The concept of using seaweeds in poultry diets has been the subject of considerable research over at least the past two decades in particular. In livestock feed, seaweeds function as sources of complex carbohydrates, with prebiotic activities and pigments and polyunsaturated fatty acids, which are known to be beneficial to animal health [9]. Multiple species of brown, green and red seaweeds, either alone or in combination, are already being commercially used in the U.S. and Canadian poultry markets. The chemical compositions of different seaweeds are indeed highly variable, being dependent on the species, time of collection and habitat, temperature and light intensity as well as nutrient concentrations in their habitat and seawater. Brown seaweeds contain a range of bioactive compounds; however, they generally have lower nutritional value than red and green seaweeds. Brown seaweeds are rich in minerals (14%–35% dry matter) and some can accumulate iodine over 30,000 times higher than that found in seawater (1500–8000 ppm vs. 0.05 ppm, respectively) [10]. Red seaweeds may be rich in proteins (10%–50% dry matter) and contain lower levels of iodine (0.03%–0.04%) [10,11]. Green seaweeds such as Sea Lettuce (*Ulva* spp.) may contain higher amounts of proteins (up to 15%)

as compared to brown seaweeds. Green algae are rich in total fiber (290–670 g/kg), with a high content of both soluble and insoluble fibers [12].

For the production of seaweed meal, the post-harvesting steps must be performed quickly in order to avoid contamination, primarily by molds. Seaweeds are usually dried and ground to fine particles (300 and 900 μm) for supplementation in poultry meals. Drying of seaweeds should not exceed 50–70 °C in order to protect the bioactivity of the functional metabolites contained therein [13,14]. The purported health benefits for the inclusion of seaweed meal in poultry diets are explained below.

2.1. Broiler Health

2.1.1. Green Seaweeds

The ubiquitous green seaweed *Ulva* spp. has been studied extensively as a substitutional feed ingredient in the diets of broiler chickens. The replacement of corn with 3.0% *U. lactuca* in the diets of male broilers, from days 12–33 post-hatch, improved the yield of breast muscle and dressing percentage and showed a numeric improvement in body weight gain (BWG) for birds fed 3% vs. 1%. These enhancements were attributed to the availability of soluble fibers and essential sulphur-containing amino acids including methionine and cysteine [15]. Inclusion of 3.0% *U. lactuca* significantly reduced abdominal fat (i.e., related to cholesterol and triglycerides) in treated birds. However, production parameters including feed intake (FI), BWG and feed conversion ratio (FCR) were the same in treated birds as in the control group. Lower inclusion levels of *U. lactuca* of up to 3% did not demonstrate toxic or anti-nutritive effects on broiler health [15]. However, the inclusion of 4% and 6% of *U. rigida*—as a prebiotic feed additive in a broiler diet was found to improve FI, FCR and mortality. In addition, blood serum total cholesterol and triglyceride levels were lower in *Ulva*-fed birds over the controls. Intestinal histo-morphology (Figure 1), including villi width, height, and length, was greater in birds fed *U. rigida* feed as compared to basal diets.

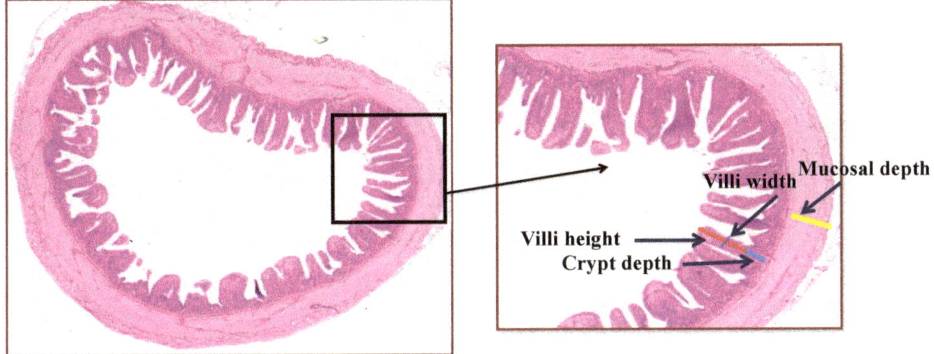

Figure 1. Histo-morphological parameters, including villi height, villi width, mucosal depth and crypt depth, for a histology section (0.5 μm thick, stained using haematoxylin and eosin staining using the procedure of Drury and Wallington (1980) and the Tissue-Tek® DRS™ (Sakura Finetek USA Inc., Torrance, CA, USA)) prepared at the ileocaecal junction region of the gastrointestinal tract (GIT) in a laying hen (Lohmann Brown Classic, 67 weeks) (Original figure by G.K.).

The prebiotic effects of *U. rigida* were similar to other prebiotic feed supplements including BIO-MOS® and inulin [16]. An increased intestinal villi length resulted in both a larger intestinal surface area and increased activity of the brush-border enzymes, leading to an increased surface area for absorption and digestive capacity [17]. Serum total cholesterol and triglyceride levels were significantly lower in *Ulva* treatments as compared to controls [17]. The differences in outcome of these studies with dietary inclusion of *Ulva* spp. may be attributed to factors such as the amount of seaweed supplemented, the purity of the seaweed, drying method, particle size, various methods of meal preparation and differences between species. All these variables should be considered in the construction of any seaweed-derived meal. However, it seems to be a common feature that levels of inclusion in the diet are generally low (up to 6%). The seaweeds are not feed replacers in their own right, but they work (perhaps synergistically) to improve bird health and resistance to disease and, therefore, help them to grow faster, with better quality when added at lower rates as supplements or prebiotics.

2.1.2. Brown Seaweeds

Brown algae are rich in functional polysaccharides such as alginates and fucoidans, which are known to have various biological activities including anti-coagulant, anti-inflammatory, anti-viral and anti-tumoral properties. These seaweed components have been evaluated as feed additives to improve broiler performance. For example, by-products of the brown seaweed *Undaria pinnatifida* have been evaluated as a dietary supplement in broiler diets. Seaweed by-products, which are components of thalli (plant components that lack differentiation into distinct parts such as stem, leaves and roots) and which do not grow from an apical point, are not consumed as food. Brown seaweed by-products, at an inclusion level of 0.5% in broiler diet, resulted in higher BWG, improved blood serum profile, immune response and a reduced mortality rate as compared to a control diet [18]. Basal diet supplementation with 100 and 200 mg/kg of a fucoxanthin extract increased catalase (CAT), superoxide dismutase (SOD) activities and glutathione (GSH) levels and decreased malondialdehyde (MDA) levels in the liver, breast and drumstick tissues. These results were taken to demonstrate that fucoxanthins could be used to regulate the antioxidant metabolism and improve the immune system of broilers [19].

It is well documented that the antioxidant status of birds plays an important role in their resistance to various infections, maintenance of health and production and reproductive performance [20]. In another study, dietary supplementation with polymannuronate (a brown seaweed derivate), at inclusion levels of 0.1%, 0.2%, 0.3% and 0.4%, altered the cecal microbiome, increased the concentration of lactic and acetic acid in the cecum and improved broiler chicken performance (i.e., average daily gain (ADG), FCR, antioxidant capacity and immune status) compared to the control diet [18]. This indicated that brown seaweed-derived compounds can improve the immune status, antioxidant capacity and performance of broiler chickens.

The addition of *Ascophyllum nodosum* (*A. nodosum*, 0.05% of feed) to broiler feed reduced the effect of prolonged heat stress while not negatively affecting growth and feed conversion, indicating that this type of feed supplementation can be used to improve bird welfare during heat stress events in poultry production [21]. Due to climatic change, meteorological events causing heat stress are of increasing occurrence. Moreover, poultry production is often carried out in regions of the globe where temperatures can reach 50 °C; the costs of cooling would be difficult to pass on with tight margins. Hence, the addition of *A. nodosum* at low inclusion levels in the diets of poultry birds can reduce the requirement (and therefore the associated cost) of cooling poultry barns, as well as the consequences of heat-associated increased mortality and lost production.

The nutritional value of various brown seaweeds of the genus *Sargassum* spp., applied in different formats including raw or thermally treated (i.e., boiled and autoclaved), were evaluated in broiler diets at 2%, 4% and 6% inclusion levels. However, the inclusion of raw or thermally treated seaweeds showed no significant effects on carcass characteristics. In contrast, the blood plasma profiles of treated birds were significantly altered, including elevated plasma high density lipoprotein (HDL) and reduced total cholesterol concentrations as compared to the control birds [22].

A recent study by Kumar (2018) demonstrated the effects of dietary supplementation of *Sargassum wightii* in broiler diets. Dried *S. wightii* powder at 1%, 2%, 3% and 4% improved BW, FI, FCR and meat quality of broilers. Dietary inclusion of 1% and 2% *Sargassum* reduced both blood plasma cholesterol and globulins and also improved total serum proteins, albumin, calcium, phosphorous and triglyceride levels in treated birds. Results from this study indicated that inclusion of 1% or 2%

Sargassum powder had the optimal supplementation effects. *Sargassum* improved dietary palatability whilst resulting in higher FI and enhanced digestibility and intestinal absorption, leading to improved BWI, as compared to controls. A higher FCR subsequently improved meat quality and carcass yield in treated birds, leading to cost efficiency. Active ingredients from *Sargassum*, including saponins, hemicelluloses, mucilage, tannins and pectin, were implicated as altering blood low density lipoprotein (LDL)-cholesterol by inhibiting bile salts [23]. Beneficial effects in broilers might also be attributed to a rich content of minerals, vitamins, long-chain fatty acids, essential amino acids, sterols and fucoidans in *S. wightii*. The degree of enhancement of broiler performance with dietary inclusion of *Sargassum* supplement can be attributed to factors such as amount supplemented, the purity of the seaweeds used and differences in seaweed meal preparation (drying and particle size). Tasco®, a branded product made simply from rapidly sun-dried *A. nodosum*, has been demonstrated as a prebiotic for broilers and can be used as an alternative to antibiotic growth promoters. Addition of Tasco® improved the growth and performance of broilers at very low inclusion levels (0.25% and 0.5%), thus increasing its cost effectiveness (and enabling the use of the term "super-prebiotic"). Tasco® displayed improvements in growth comparable to the positive control inulin (a standard prebiotic derived from chicory) and the antibiotic virginiamycin. Tasco® showed effectiveness in the lower gastrointestinal tract (GIT) by altering the pH of the intestine, intestinal histo-morphology and bursa and cecal relative weights, indicating its fermentation in the lower GIT by beneficial microflora [24].

2.1.3. Red Seaweeds

Red seaweeds including *Chondrus crispus* (Irish moss) and *Palmaria palmata* (dulse) have high nutritive values and have been considered to be highly palatable to poultry and ruminant animals. Dried red seaweeds, e.g., *Polysiphonia* spp. (up to 3%), were shown to serve as an intermediate source of protein to growing broiler chicks. *Polysiphonia* contains elevated levels of proteins (i.e., 32.4%) and minerals as required by rapidly growing poultry. However, inclusion had no significant effect on overall growth performance [14]. Calcified seaweeds can function as an alternative source of dietary calcium, which resulted in increased bone health and reduced leg weakness and lameness as compared to calcium obtained from limestone. The inclusion of the calcareous marine algae (CMA, at 0.45%, 0.6%, 0.75% and 0.9%) reduced both feed intake and bird growth, with a negative impact on bone strength, since tibia ash and phosphorus levels were lower in birds fed with calcium (0.9%) from CMA. However, ileal calcium digestibility had a linear increase in birds fed with 0.45% CMA [25]. Higher dietary calcium from limestone decreased phosphorous digestibility in broilers, which was shown to be improved by the inclusion of lower concentrations of calcified seaweeds [26,27]. Inclusion of *P. palmata* (1.8%) in broiler diets improved body weight and increased beneficial bacteria (e.g., *Lactobacillus*) in the ileum, serum IgA and ileal villus width, height and surface area [28]. Feed supplementation with *Kappaphycus alvarezii* (AF-KWP) improved body weight gain and feed intake and increased the haemagglutination (HA) titre and cell-mediated immunity (CMI) levels. Inclusion of 1.25% AF-KWP in a broiler diet positively affected performance, immunity and breast yield in broiler chickens [29]. Dietary inclusion of the commercial red seaweed, dulse (*P. palmata*) (Organic Whole Leaf-Dulse, Vitaminsea®) at 0.15% showed beneficial effects on growth performance, cooking loss, drip loss, diarrhea score and the fecal microbiome (i.e., it significantly reduced the relative abundance of pathogenic bacteria including *E. coli* and enhanced beneficial bacteria including *Lactobacillus*) [30]. A similar response of decreased "shedding" of intestinal *E. coli* O157:H7 was observed in beef cattle when sun-dried *Ascophyllum nodosum* seaweed (i.e., Tasco-14™) was added to their diets. Administration of Tasco-14™ at a level of 20 g/kg diet for 7 days was effective at lowering both the duration and intensity of *E. coli* O157:H7 fecal shedding by cattle [31]. These beneficial effects can be due to the presence of dietary sulphated polysaccharides in seaweeds. Similarly, economically viable seaweeds can be administered to pre-slaughter chickens in order to evaluate reductions in the shedding of pathogenic bacteria.

2.2. Health of Laying Chickens

2.2.1. Green Seaweeds

Ulva prolifera and *Cladophora* sp. are enriched in micro-elements including Cu(II), Zn(II), Co(II), Mn(II) and Cr(III), and improved the average body weight of treated laying hens, resulting in a higher average egg weight and eggshell thickness vs. the controls. Laying hen diets supplemented with seaweeds enriched with micro-elements also resulted in higher microelement transfer to eggs and enhanced the colour of yolk [32]. Inclusion of *U. prolifera* at 1%, 2% and 3% improved immune function, egg production and egg quality (egg weight, shell thickness and yolk colour) whilst also reducing and/or improving the feed conversion ratio and yolk cholesterol. In addition, the abundance of beneficial microbes, including *Bifidobacterium* and *Lactobacillus,* was significantly increased in the feces of laying hens as compared to control groups, indicating better animal health [33]. Ulvan (i.e., a sulphated polysaccharide extract from the green seaweed *Ulva*), when added to diets of brown laying hens at 0.5%, 0.8% and 1%, enhanced the function of the small intestine and regulated the digestive system, resulting in improved egg production, egg weight and FCR. This could be of great benefit to poultry farmers, as ulvan did not increase the feed intake but enhanced the egg weight [34]. It is possible that other sulphated polysaccharides (from brown and red seaweeds) have similar functionalities, but this remains to be investigated.

2.2.2. Brown Seaweeds

Incorporation of 10% *Macrocystis pyrifera* (giant kelp) in meal enriched with n-3 FA from fish oil in the diets of 35-week-old Leghorn hens effectively increased egg n-3 FA content, albumen height and yolk colour [35]. Sensory evaluation of these eggs revealed that flavour was not affected by the treatment. In another study, the effects of different concentrations of brown algae (BMA, *Sargassum dentifebium*, 3% and 6%) prepared using different methods (i.e., sun-dried, SBMA; boiled, BBMA; autoclaved, ABMA) on egg profiles were reported. Inclusion of 3% or 6% BMA meal in the laying hen diet significantly reduced plasma cholesterol, as well as yolk cholesterol and triglycerides, whilst also improving the total palmitic acid, carotene, lutein and zeaxanthin levels in eggs [36]. By-products from *Undaria pinnatifida* and *Hizikia fusiformis* (0.5%) were shown to improve egg laying performance and relative organ weights, particularly the liver and cecum, over those of the control group. This study demonstrated that supplementation with seaweed by-products resulted in superior bird health [37]. Dietary supplementation by the commercial brown seaweed *Ascophyllum nodosum*, trademark name Tasco® (at 0.25% and 0.5%), significantly enhanced egg weight, shell weight and yolk colour in eggs from Lohmann Lite hens (age = 70 weeks). Hens fed a diet with 0.25% Tasco® had significantly larger eggs and shell weight as compared to hens fed 0.5% and the control diets, indicating that lower inclusion levels of Tasco® enhanced both productivity and economic efficiency in poultry production [38].

2.2.3. Red Seaweeds

Inclusion of red seaweeds, e.g., *Chondrus crispus* (CC, 1%) and *Sarcodiotheca gaudichaudii* (SG, 2%), in standard poultry diets improved FCR and egg quality parameters. The SG and CC groups showed greater height and surface area of villi as compared to the control birds. Seaweed supplementation also increased the abundance of beneficial gut bacteria, e.g., *Bifidobacterium longum* (4-14-fold) and *Streptococcus salivarius* (4-15-fold), and reduced the prevalence of *Clostridium perfringens*. Additionally, the concentration of short chain fatty acids, including acetic acid, propionic acid, n-butyric acid and i-butyric acid, were significantly higher for both CC and SG treatments [39]. *Gracilariopsis persica* meal fed at 50 gm/kg (5%) significantly lowered the levels of cholesterol and malondialdehyde in egg yolk vs. control birds [40]. Dietary inclusion of the red seaweed *Kappaphycus alvarezii* (1.5%) significantly reduced egg laying age and improved production parameters and egg quality traits (egg production, egg weight, shell thickness) in laying hens [41]. Taken together, these studies suggest that dietary supplementation with selected red seaweeds as a potential prebiotic source is associated with improved performance, egg quality and overall gut health in laying hens.

3. Novel Formulations of Seaweeds for Poultry Health

Processing/modification of seaweeds can improve the bioavailability of their active components in poultry feed, thereby impacting both the digestibility and performance of chickens. The following strategies have been reported to improve feed efficiency and palatability of seaweeds in livestock feed. The drying and pre-treatment phases are very important for the maintenance of seaweed quality during storage.

3.1. Mechanical Approach

Feed processing methods include drying, cooling, pelleting, cooking, vacuum coating, steam exploding and extruding. These processes are utilized in order to be cost-effective (provide target nutrient at least/best cost) and to improve digestibility and feed efficiency in chickens. Feed technology has advanced from basic mixing of a mash feed to more innovative preparations involving physical and hydrothermal processing operations. In a commercial setting, feed processing includes single or multiple processing of feedstuffs in order to meet objectives. Poultry diets are manufactured using a combination of technologies such as grinding with hammer and/or roller mills, along with hydrothermal processing including pelleting, expansion or extrusion. The major advantages of feed processing are the improved availability of nutrients, destruction of inhibitors and toxins and reduction of feed wastage [42].

3.1.1. Size Reduction

The particle size of feed in the diet plays an important role in the development of the digestive tract and regulation of feed intake by birds. Birds consuming larger particle-size feed develop larger, more muscular gizzards and longer intestines. In addition, larger feed particles require more time for breakdown in the gizzard and intestine (Figure 2), resulting in longer microvilli and an increase in surface area, thus positively affecting digestibility and absorption [43]. Raw seaweeds are mainly dried and ground to pass a 0.3–1.0 mm mesh screen using a Wiley mill or grinders. Size reduction by grinding is the most economical method utilized in poultry feed preparation.

3.1.2. Extrusion

Feed extrusion is a combination of heat, shear and compressional forces utilized to produce strongly bonded and porous pellets. Feed prepared by extrusion with *Chondrus crispus* (0.5%–3%) had no effect on either egg quality or production parameters, indicating that minimal processing by simple grinding was satisfactory compared to the added cost of mechanical processing [38]. Birds fed on the 3% *Chondrus* diet produced larger and heavier eggs, but no other significant differences were observed in 3% vs. 1% and/or 2% inclusion. Hence, the recommended levels of *Chondrus crispus* supplementation to laying hen feed were 1%–2%, which is also more cost-effective. On the other hand, the recommended level of Tasco® (air dried, brown seaweed *Ascophyllum nodosum*) in laying hen feed was 0.25%, which is predicted to be the most cost-effective.

3.2. Additive/Synergistic Approaches

3.2.1. Biological Treatment by Fermentation

In poultry, feed processing by fermentation can produce functional feeds which are formulated to improve the gut microbiome, health and performance. Major functional ingredients introduced by this treatment include higher numbers of lactic acid bacteria, a reduced pH and high concentrations of organic acids. These features protect the feed from microbial contamination during storage [45,46]. Fermentation enhances the antioxidant, anti-coagulant and anti-inflammatory effects of seaweeds, and it also increases the stability of feed during storage. In addition, beneficial microorganisms used in microbial fermentation can have probiotic effects on poultry performance. Thus, fermentation of seaweeds by probiotic bacterial strains could introduce synergistic effects [47]. Feed supplementation with fermented brown seaweeds, e.g., *Undaria pinnatifida* (0.5%), improved the weight gain, feed: gain ratio and immune status of broiler chickens as compared to controls. Blood serum profiles including glutamic pyruvate transaminase (GPT) and concentrations of immunoglobulins (IgA and IgM) were significantly higher in fermented seaweed treatments than their controls. However, the IgG titers were decreased as compared to controls [18]. These observations indicated that fermentation of these dietary seaweeds by *Bacillus subtilis* improved the growth performance and immune profile in broilers [18]. Conversely, the same authors concluded that supplementation with fermented seaweeds had no beneficial effect on laying hen performance. Bacterial fermentation was proposed to result in depletion of oligosaccharides, which would decrease the positive supplementation effect of seaweeds as prebiotics [37]. Altering the conditions/environment during fermentation by adding acidifiers, e.g., organic acids, concentrated starter lactic acid bacteria (LAB) strains or enzymes, can speed up the fermentation processes as well as improve the functional characteristics and palatability of the final product [48]. With improved palatability, feed intake by chickens can be increased, leading to positive effects on growth performance, gut microbiome and morphology. Clearly, whilst promising, further work is required in this fledgling area of application.

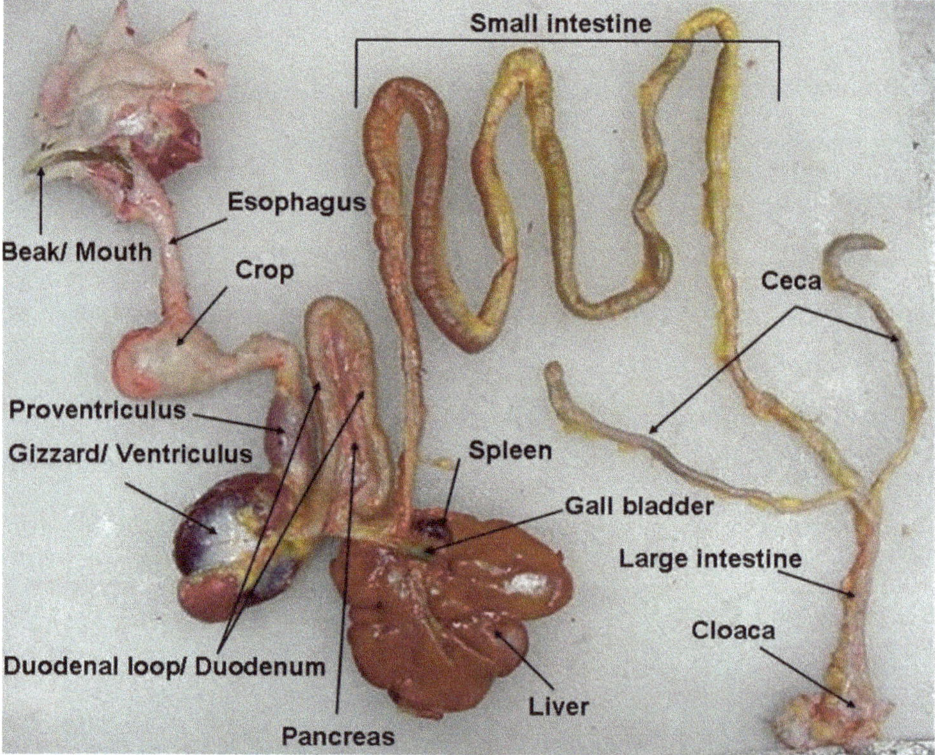

Figure 2. Diagrammatic representation of different parts of the digestive organs of a typical chicken. (adapted from [44]).

3.2.2. Botanical Blends

Seaweeds can be used in combination with other natural bioactives including essential oils, green tea and anti-microbial peptides. Such combined products function as botanical alternatives to chemicals and antibiotics, for use in certifiably organic poultry production. For example, a combination of the green alga *Ulva* (cited as *Enteromorpha)* (10%) and sardine oil (2%) provided a source of antioxidants in the laying hen feed and also enhanced the DHA (docosahexaenoic acid) levels in eggs [49]. Similarly, the brown algae *Macrocystis pyrifera* and *Sargassum sinicola* (administered at 10%) enhanced the EPA (eicosapentaenoic acid) content of the eggs. In general, eggs can be conserved at 4 °C; however, the concentration of fatty acids declines with increasing storage time. The inclusion of sardine oil in the diets of laying hens was observed to increase the n-3 polyunsaturated fatty acids (PUFAs) content of eggs; however, n-3 PUFAs are more sensitive to oxidation, resulting in rancidity in the final product and leading to reduced shelf life. [49]. Thus, it can be inferred that supplementation of laying hen diets by selected seaweeds, in combination with essential oils, can produce enriched eggs with improved shelf life. In another study, co-supplementation of a laying hen diet with green tea (at 0.1% and 0.2%), combined with red and green seaweeds (also at 0.1% and 0.2%), led to improved egg production, egg quality and physiological and immunological performance of late phase laying hens [50]. Co-supplementation with dietary *Laminaria japonica* (brown seaweed) powder (3%) and anti-microbial peptide (300 mg cecropin/kg, 0.03%) significantly improved growth performance (i.e., increased FCR) and immune function (serum Newcastle disease antibody titers and lymphocyte numbers) in broilers. In addition, the same co-supplementation reduced *E.coli* and increased *Lactobacillus* levels in the cecum of broiler chickens, indicating the potential use of *L. japonica* powder and cecropin as an alternative to antibiotics in broiler production [51].

Algae-based antioxidant supplements containing selenium yeast (EconomasE®, Alltech Inc., Nicholasville, KY, USA), when added to broiler diets, significantly improved the meat quality attributes, including water holding capacity, tenderness, colour and pH; thus, EconomasE® can be used as a nutrient supplement in broiler diets [52]. Contamination by mycotoxins as a result of the spoilage of the feed results in undesirable health effects and a decline in the rate of egg production, with adverse economic effects. Hence, control of fungal development and mycotoxin production are critical for feed and animal producers. Addition of EconomasE® (2 gm/Kg, 0.2%) to mycotoxin-contaminated corn diet was demonstrated to partially improve the production performance (FCR) in broiler chickens [53]. Mycotoxins such as aflatoxins (AF), ochratoxin A (OTA), fumonisins (FUM), deoxynivalenol (DON) and T-2 toxin adversely affect the health and productivity of poultry. It was suggested, but not fully tested, that Tasco® or dried *Ascophyllum* meal could have mycotoxin-binding effects [53].

3.2.3. Algal Clay

Seaweed containing feed supplement MFeed+® Olmix, Brehan, France, has been developed by associating algal extracts (*Ulva* sp. and *Solieria chordalis*) with clay (bentonite) for use in livestock diets. Clays contain layered mineral materials organized in a succession of aluminum and silica-based sheets. Some seaweeds contain high levels of trace mineral ions (e.g., iron, zinc, copper, titanium) that can function as co-factors for enzymes and so improve their activities. Moreover, clay has been shown to slow down the transit time of feed in the intestine, thereby increasing digestion and resulting in better feed efficiency and nutrient uptake [54]. Such effects of clay have also been established in pigs [55].

In broilers, supplementation with a clay mix improved the weight gain and feed efficiency (feed intake and growth). Seaweeds in the clay mix introduced trace mineral ions into the diet, which improved the activity of some digestive enzymes and resulted in increased growth performance of the broilers fed on the supplemented diet. Algal clay (0.1%) can be incorporated to reduce the cost of feed while still maintaining a productive performance in broilers [56]. An algal clay-based product (i.e., MT.X+®, Olmix product) added to the diet prevented the negative effects of mycotoxin contamination on performance and productivity in broilers, at both experimental and commercial scales. The product improved the production efficiency factor by 10% and the return on feed cost by 36% when compared with the control [57].

In 2016, the European Food Safety Authority (EFSA) Panel on Additives and Products or Substances Used in Animal Feed (FEEDAP Panel) delivered a scientific opinion on the safety and efficacy of an algal/clay mix for animal consumption [58]. The panel concluded that the additive product, composed of feed-grade bentonite and selected seaweeds, was considered safe for livestock (e.g., piglets, cows and chickens) consumption, at a maximum recommended dose of 124 mg/kg (0.0124%) of complete feed. The additive product is considered non-genotoxic (bentonite is not absorbed from the gut lumen and the seaweeds were shown to have beneficial effects in humans) and were safe for animal nutrition and for consumers [58].

4. Anti-Bacterial and Anti-Viral Effects of Various Seaweeds on Disease in Poultry Production

The use of seaweeds as anti-infective agents in commercial livestock production has gained interest due to an increase in antibiotic-resistant bacterial strains and increasing consumer concerns regarding drug residues in animal meat. Seaweeds are a rich source of dietary fiber, minerals, vitamins, proteins, phlorotannins and carotenoids [59]. Seaweeds in poultry diets enhance gut microbiota, as the algal biomass remains mostly undigested in the lower GIT, and therefore act as substrates for bacterial fermentation [60]. Red and brown seaweeds have prebiotic-like properties that alter the metabolic activities of beneficial microflora and reduce the prevalence of pathogenic bacteria [61]. Moreover, a carbohydrate fraction extracted from the red seaweed *Gracilaria persica* exhibited direct anti-microbial effects against six bacterial pathogens including *Staphylococcus aureus*, *E. coli*, *Methicillin-resistant Staphylococcus aureus* (MRSA), *Salmonella typhimurium*, *Pseudomonas aeruginosa* and *Aeromonas hydrophila* and induced a humoral-immune response against sheep red blood cells (SRBC) [62]. Likewise, phlorotannin extracts isolated from two brown seaweeds *A. nodosum* and *Fucus serratus* were effective at killing three foodborne pathogens, *E. coli* O157, *Salmonella agona*, and *Streptococcus suis*, without negatively affecting the pig intestinal cells (in vitro) [63]. Water extracts of the red seaweeds *Gelidium latifolium*, *Hypnea musciformis*, *Jania rubens*, *Jania* spp. and *Laurencia obtusa* showed significant in vitro anti-microbial activities against pathogenic, Gram-negative bacteria, including *E. coli*, *Klebsiella* spp. and *P. aeruginosa* [64]. Moreover, sulphated galactans and carrageenans from an aqueous extract of the calcareous red alga *Corallina* sp. possessed bactericidal activity against pathogenic Gram-positive bacteria including *Enterococcus faecalis* and *Staphylococcus epidermidis*. Taken together, these studies indicate that the organic and polysaccharide fractions of selected red seaweeds can function directly as anti-microbial components in poultry diets. In addition, seaweed polysaccharides such as carrageenans, sulphated proteoglycans, and dextran sulphates have been reported to possess a broad spectrum of anti-viral activities [65–69]. A number of sulphated polysaccharides are potent inhibitors of paramyxoviruses, including parainfluenza virus, respiratory syncytial virus, mumps virus, measles virus, Newcastle disease virus (NDV) and distemper canine virus [68,70–73]. The aqueous extracts of the red alga *Schizymenia dubyi* with the highest sulphate content were effective in inhibiting HSV-1 replication at an EC_{50} = 2.5–80 µg/mL without cytotoxic effects. Methanolic and 2,3,6-tribromo4,5-dihydroxybenzyl methyl (TDB) ether extracts isolated from the red alga *Symphyocladia latiuscula* exhibited anti-viral activities against wild type HSV-1 and acyclovir (ACV) resistant-HSV-1 (IC_{50} values of 5.48, and 4.81 µg /mL, respectively). Daily oral administration of the methanolic and TDB extracts delayed the appearance of lesions in infected mice, without toxicity [74]. Similarly, lambda-carrageenans from the red seaweed *Gigartina skottsbergii* (Gigartinaceae) displayed anti-viral activity against animal viruses belonging to the *Alphaherpesvirinae* sub-family BoHV-1 (bovine herpesvirus type 1) strain Cooper and SuHV-1 (suid herpes virus type 1) strain Bartha [75]. These results indicated that seaweed components, primarily polysaccharides, have potential as anti-viral agents in poultry diets. Table 1 describes recent studies of the use of various seaweeds as anti-microbial (bacterial, viral, plasmodial, etc.) in poultry diets in order to improve animal health and performance.

Table 1. Use of seaweeds (SW) as anti-microbials in poultry diseases.

Macroalgae B = Brown R = Red G = Green	Level of Inclusion in Feed	Anti-Microbial Response/Poultry Disease	Reference
Laminaria japonica (LJP) *(B)* and anti-microbial peptide cecropin	*Laminaria japonica* LJP:1%, 3% and 5%; Cecropin: 0.03%	Anti-bacterial and anti-viral activities were observed with dietary supplementation of broiler diets with LJP + cecropin, which increased feed conversion ratio (FCR), and serum Newcastle disease antibody titers and lymphocyte numbers. In addition, birds fed with LJP showed significant inhibition of *E. coli* counts and increase in *Lactobacillus* counts in ceca.	[51]
Ascophyllum nodosum (B)	0.05% and 0.1%	Anti-bacterial activity. *A. nodosum* reduced *C. jejuni* counts in the caecum of chicks (10 days old), at both concentrations, but decreased the growth parameters (disruptive effect on gut morphology in ileum). Significant increases in the expression of tight-junction genes OCLN and CLND-1 alongside increases in MUC2 and CCND1 expression.	[76]
Chondrus crispus and *Sarcodiotheca gaudichaudii* (R)	2% and 4%	Anti-bacterial activity. The incorporation of SW in the diets of Lohmann Lite laying hens reduced the negative effects of *Salmonella enteritidis* (SE) infection on body weight and egg production.	[77]
Grateloupia filicina, (R) *Ulva pertusa* (G) and *Sargassum qingdaoense* (B)	In vitro, 20–500 mg/mL of sulphated polysaccharides SPs; in vivo mouse model, 0.001% and 0.005% of SPs per day	Anti-viral activity. Sulphated polysaccharide extracts from all three species showed immune-modulatory activities, both in vitro and in vivo; *S. qingdaoense* showed the best activity. All three SPs significantly inhibited the activity of activated AIV (H9N2 subtype) in vitro and inactivated avian influenza virus (AIV) in vivo. Sulphated polysaccharides from *G. filicina* showed the strongest anti-AIV response.	[78]
Ulva clathrata (G) and fucoidan	In vitro 0.1–1000 µg/mL	Anti-viral activity. The ulvan and fucoidan extracts inhibited Newcastle disease virus (NDV) in vitro and showed no cytotoxicity at effective concentrations. Ulvan inhibited viral fusion by interacting with the intact F0 protein. Ulvan exhibited better anti-cell–cell spread activity than fucoidans, but a combination showed more potent (synergistic) responses.	[79]
Highly soluble calcified seaweed (HSC) (R)	0.6% and 0.9%	Anti-bacterial activity. Broilers fed HSC diets had significantly higher feed conversion/total weight than control birds. Lower dietary Ca (0.6% vs. 0.9%) showed lower mortality associated with necrotic enteritis (NE) as compared to higher dose (0.9%) and on bird performance.	[80]
Chaetomorpha antennina (G) in combination with mangrove species *Aegiceras corniculatum* (land plant)	In vitro: 0.5, 1.0 and 1.5 mg/mL In vivo rat model: 0.02% per day	Parasite inhibition. The extract mixture showed 60% suppression of parasitaemia against *Plasmodium falciparum* at 1.5 mg/mL. Anti-plasmodial activity (50%) against *Plasmodium berghei* was observed in vivo.	[81]

4.1. Mechanism of Anti-Microbial Activity of Seaweeds

4.1.1. Anti-Bacterial Mode of Action

Seaweeds are continuously exposed to a range of abiotic stresses such as desiccation, sunlight, osmotic stress and extreme temperatures, as well as pathogenic microbes. In response, seaweeds have developed protective mechanisms in order to combat and survive these stressful conditions [82]. They produce an array of unique bioactive compounds, including sulphated polysaccharides, organic acids, pigments and phenolic compounds, which are responsible for a range of functionalities, such as antioxidant, anti-microbial and anti-viral activities. For example, phenolic compounds exhibit anti-microbial activity by permeabilizing the bacterial cell wall and releasing the intracellular contents [83]. Other mechanisms of action of phenolic compounds against bacteria include interference with nutrient uptake, impairment of protein and nucleic acid synthesis and disruption of electron transport chains [83]. On the other hand, seaweed-derived polysaccharides can elicit defense responses in the host which are similar to pathogen recognition (PAMP triggered immunity) [84]. Red seaweed-derived polysaccharides also exhibit anti-microbial activity because of their affinities towards surface appendages of the bacteria. Anti-microbial activities of red seaweeds and their extracted compounds on the poultry pathogen *Salmonella* Enteritidis (SE) have been linked with the down-regulation of virulence factors, restricted motility and flagellar functions and also direct the blockage of bacterial quorum sensing (Figure 3). Quorum sensing molecules such as auto-inducers (acylated homoserine lactones, AHL) have been shown to facilitate virulence, motility and biofilm formation in bacterial pathogens including *Salmonella* [85,86]. Previous studies have shown that some red seaweeds contain quorum sensing inhibitors, such as brominated furanones, which are capable of inhibiting bacterial biofilm formation and the regulation of flagellar and virulence genes, resulting in bacterial growth inhibition [87,88].

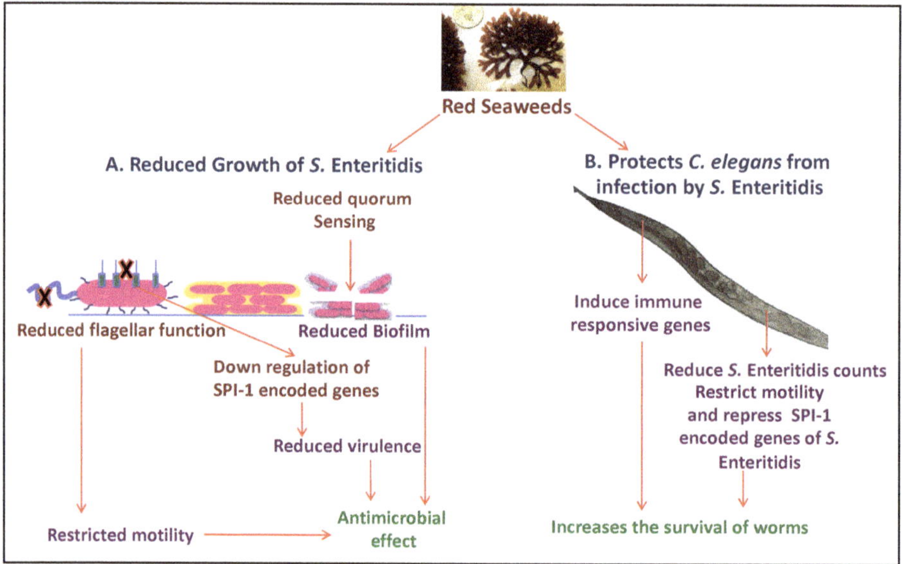

Figure 3. Illustration of modes of actions of selected red seaweeds, *Chondrus crispus* (CC) and *Sarcodiotheca gaudichaudii* (SG). A. Direct anti-microbial effect: red algal extracts inhibited growth, motility, biofilm formation and quorum sensing mechanisms in *Salmonella* Enteritidis. B. Protection of *Caenorhabditis elegans* from infection by *S.* Enteritidis: both red seaweed extracts increased the survival of infected worms by reducing *S.* Enteritidis colonization in *Caenorhabditis elegans* and enhancing the immune response of the worms (redesigned from [89]).

Yeast cell wall-derived mannan polysaccharides have been shown to deactivate Gram-negative pathogens such as *E. coli* and *Salmonella* by competitively binding to surface appendages such as the fimbriae and flagella. In the intestine, the adherence of pathogens to polysaccharides such as mannans reduces their ability to attach to epithelial cells, which results in the complete clearance of pathogenic bacteria from the gut without colonization [90]. Seaweed polysaccharides, with their ionic properties, have been shown to exhibit anti-bacterial activity against Gram-negative bacteria. For example, the anti-microbial activity of alginic acid (from brown seaweeds) against *E. coli* has been attributed to its polyanionic nature [91].

Another mechanism by which red seaweeds reduce the colonization of *S.* Enteritidis in the ceca of laying hens has been attributed to the attenuation of the virulence factors of SE (Figure 4). One study demonstrated that *C. crispus* (CC) and *S. gaudichaudii* (SG) water extracts reduced the relative expression of virulence factors of SE in vitro and decreased the colony count of SE in the intestine of *C. elegans*. Water extracts of seaweeds (CC and SG) significantly increased the survival of *C. elegans* infected with SE and reduced the accumulation of SE in *C. elegans* gut. A decrease in the colonization of SE in *C. elegans* was likely due to (i) a significant reduction in expression of virulence-associated genes of SE; (ii) reduced ability of bacteria to attach to the surface of the intestinal epithelium of *C. elegans*; (iii) induced immune response related genes of infected *C. elegans*. The modes of action of these red seaweeds to reduce *Salmonella* colonization in the model organism (*C. elegans*) were also effective when added to the feed of laying hens. The virulence factors of *S.* Enteritidis, which are known to be essential for the colonization of the intestinal tract in *C. elegans*, can be critical for SE colonization in poultry [92].

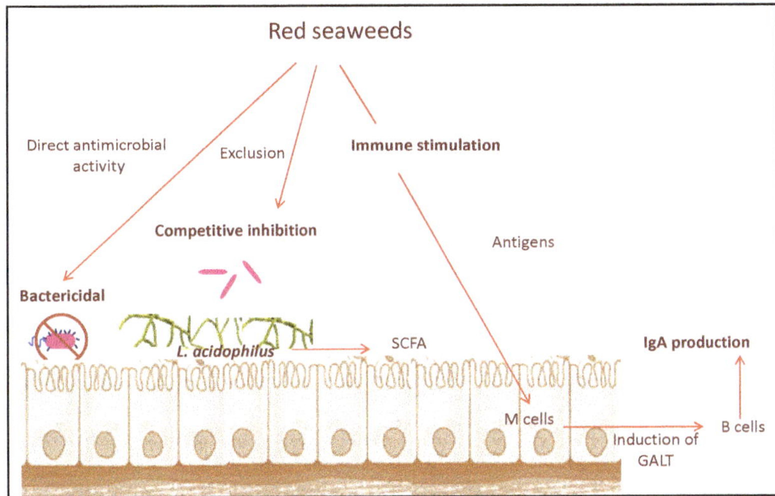

Figure 4. Dietary inclusion of red seaweeds in laying hen diets was observed to suppress the negative effect of SE on laying hen growth and performance. Various mechanisms included direct inhibition (bactericidal) of bacterial colonization of feces and ceca, competitive exclusion (i.e., reduction of *E. coli* titers) by beneficial bacteria such as *Lactobacillus acidophilus* and immune stimulation. SCFA: short chain fatty acid; GALT: gut-associated lymphoid tissue (Original figure by G.K.).

4.1.2. Anti-Viral Modes of Action

Polysaccharides and other bioactive functional molecules in seaweeds display anti-viral activity against a range of viruses by interfering with different stages of viral attachment, penetration and infection (Figure 5). Seaweed polysaccharides, such as carrageenans and galactans from red seaweeds, target viral attachment stages by either directly interacting with the virion or mimicking the binding of virus associated proteins (VAP) to the respective receptors [75,93]. Moreover, marine polysaccharides can also block the allosteric processing of the viral capsid during the internalization process and uncoating of the virus. For example, carrageenans inhibit viral attachment as well as its internalization and uncoating; ulvans inhibited fusion of Newcastle disease virus by blocking the cleavage of intact protein F0 into the mature form [79]; fucoidans inhibited viral infection by direct interaction with envelope glycoproteins [71]. Seaweed polysaccharides can also improve the host anti-viral immune response; for example, fucoidan can stimulate both specific and non-specific responses such as the activation of NK cells, maturation of dendritic cells (DCs) and activity of cytotoxic lymphocytes, as well as the ability to produce antigen-specific antibodies and memory T cells under in vitro and in vivo conditions [94].

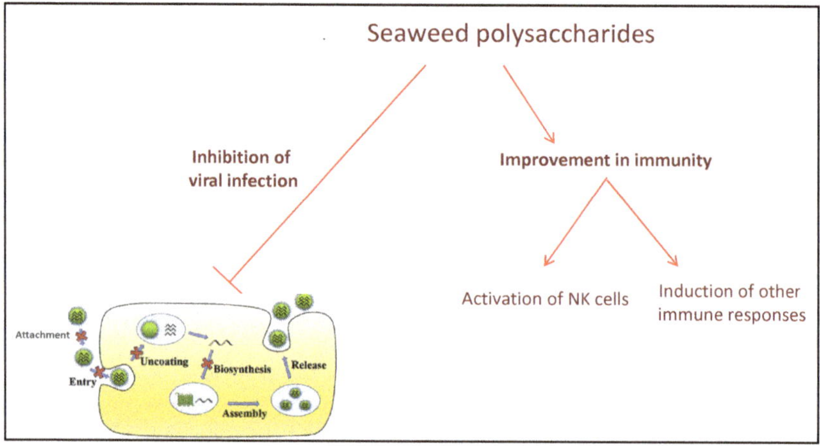

Figure 5. Mechanisms of anti-viral inhibition by seaweed polysaccharides. Seaweed polysaccharides display anti-viral activity against a range of viruses by interfering with different stages of viral attachment and replication as well as by improving host immunity (redesigned from [95]).

5. Use of Prebiotics in Poultry Production

The term prebiotics is defined as "a non-digestible food ingredient that affects the host by selectively stimulating the growth and/or activity of one, or a limited number of bacteria, in the colon" [96–98]. From a food safety perspective, prebiotics function as preventative agents which can modulate gastrointestinal microbiota in order to benefit the host and serve as a barrier to pathogen colonization. In poultry, prebiotics can induce a direct effect on birds by priming the host immune system or an indirect effect by modulating the compositing and fermentation profile of the gastrointestinal microbes [99]. Seaweeds must satisfy a number of criteria in order to be considered a prebiotic source:

(1)   They should resist digestion by acid and enzymatic hydrolysis in the upper gastrointestinal tract (GIT).
(2)   They must have a selective function as a substrate for the growth of beneficial bacteria.
(3)   They must be capable of altering the profile of the microflora.
(4)   They must induce beneficial effects that boost the host immune system and overall health.

Seaweeds and their bioactive compounds, such as polysaccharides and phenolics, exhibit these characteristics and can be considered prebiotic dietary supplements with gut health benefits. In poultry, prebiotics have been shown to improve gastrointestinal health by providing a substrate for beneficial bacteria within the gut microbiota of chickens [100]. The mode of action of most of prebiotics is by one or more of the following mechanisms: lactic acid production, inhibiting/preventing colonization of pathogens, modifying metabolic activity of normal intestinal flora and stimulation of the immune system [101]. Major beneficial probiotic bacteria present in the gut microbiome of chickens include *Bifidobacteria, Lactobacillus, Ruminococcus* and *Streptococcus*. These bacteria, which are present in the small intestine, utilize non-digestible polysaccharides and fibers for energy [102]. These beneficial bacteria can utilize seaweed polysaccharides and dietary fibers for energy and modulate the population of disease-causing bacteria in order to improve metabolism (Figure 6).

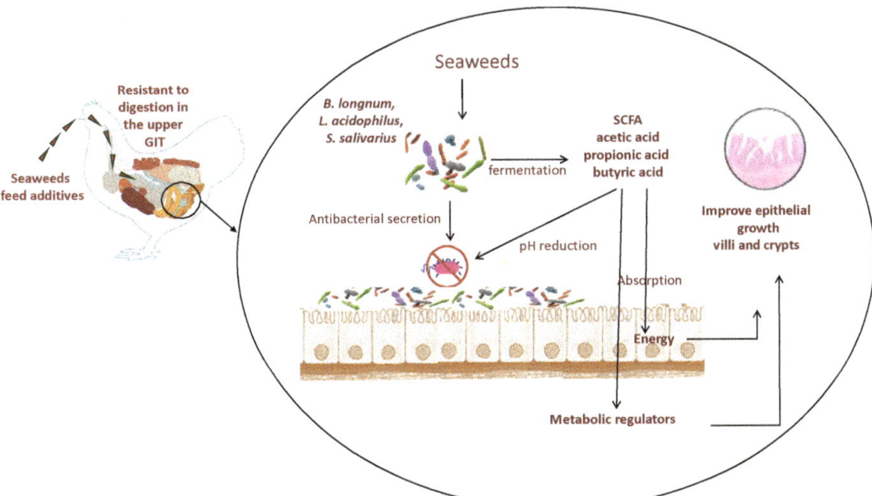

Figure 6. Modes of action of selected seaweeds as prebiotics for poultry health. Seaweeds are resistant to digestion in the upper gastrointestinal tract of chickens. After entering the colon, they are selectively fermented by beneficial microbiota, resulting in their increased numbers as well as the reduction of pathogenic bacteria by competitive exclusion. Beneficial microbes are known to produce short chain fatty acids and secrete anti-microbial peptides such as bacteriocins whilst also helping in the differentiation and proliferation of enterocytes, all of which improves epithelial growth in addition to exhibiting immunomodulatory effects (redesigned from [89]).

6. Effect of Selected Seaweeds on the Gut Microbiome

The gastrointestinal tract (GIT) of chickens possesses a diverse bacterial population which varies significantly from the proximal to distal segments. Bacterial abundance in a specific section of the GIT depends on their affinity to either enterocytes or to the mucus layer, tolerance to the GIT environment and also their resistance to the host immune system. Additional factors including the rate of passage of digesta, pH, nutrient digestibility and bioavailability and the presence of anti-microbial peptides can modulate bacterial diversity in each segment of the GIT [103]. Dietary fibers and carbohydrates present in some seaweeds enhance the growth of certain beneficial bacteria, which leads to a cascade of biological functions which then impart beneficial effects on the health and growth of the host. Gut microbial fermentation of seaweed components, and their effects on the microbiome and metabolomics, are presented in Table 2.

Various seaweed polysaccharides, including ulvans and mannans from green, fucoidans and laminarans from brown and carrageenans from red seaweeds have been associated with a range of health-promoting effects, such as prebiotic, anti-bacterial, anti-inflammatory and antioxidant functionalities. These polysaccharides are neither digested nor absorbed by the host, but they serve as a substrate for bacterial fermentation in the colon and thus impart beneficial effects on both animal and human health [104].

Table 2. Summary of prebiotic effects of different seaweed dietary fibers (in vitro and in vivo), including modulation of the gut microbiome and fermentation response.

Seaweed Source Red (R), Brown (B) or Green (G)	Component	Type of Study/Level of Inclusion	Microbiome Modulation	Metabolome Modulation	Other Responses	Reference
Palmaria palmata (R)	whole seaweed	0.6%, 1.2%, 1.8%, 2.4% and 3%. In vitro in broiler chickens	Method: Microbiology culture techniques, 16SRNA amplicon sequencing. ↑*Bifidobacterium* ↑ *Lactobacillus (ileum)* ↓ *Clostridium perfringens*	Not measured	Increasing trend in the size of villus height, width, villus surface area and mucosal depth ↑ Plasma immunoglobulin (IgA and IgG) Best response: 1.8%	[28]
Chondrus crispus and *Sarcodiotheca gaudichaudii* (R)	whole seaweed	0.5%,1% and 2%	Method: Real-time PCR. ↑*Bifidobacterium longum,* ↑*Lactobacillus acidophilus* ↑*Streptococcus salivarius* ↓*Clostridium perfringens*	↑Acetic, ↑propionic acid, ↑*n*-butyric acid ↑*i*-butyric	Increase in the size of villus height, width, villus surface area and mucosal depth Increase in ceca weight	[39]
Chondrus crispus and *Sarcodiotheca gaudichaudii* (R)	whole seaweeds	2% and 4%	Method: Microbiology culture techniques, 16SRNA amplicon sequencing. ↑ Firmicutes and Bacteroidetes ↑*Bifidobacterium longum,* ↑*Lactobacillus acidophilus,* ↑*Streptococcus salivarius* ↓ *Clostridium perfringens* ↓ *Salmonella* Enteritidis	↑ Propionic acid	↑ Plasma immunoglobulin (IgA and IgG)	[77]
Chondrus crispus (R)	whole seaweeds	0.5 and 2.5% In vivo mouse model	Method: 16S rRNA sequencing-based Phylochip array of fecal samples. ↑ *Bifidobacterium breve* ↓ *Clostridium septicum* and *Streptococcus pneumonia*	↑Acetic, ↑propionic and ↑butyric acids in faecal samples	Improvements in proximal colon histo-morphology ↑ Plasma immunoglobulin (IgA and IgG)	[105]
Gracilaria spp. (R), *Gelidium sesquipidale* (R) and *Ascophyllum nodosum* (B)	agar and alginate	Anaerobic fermentation	Method: Fluorescent in-situ hybridization. ↑Bifidogenic effect	↑Acetic, propionic	Not measured	[106]

Table 2. *Cont.*

Red seaweed	neo-agaro-oligosaccharides (NAOS) from enzymatic hydrolysis of agarose	2.5% and 5% In vivo mouse model	Method: Microbiology culture techniques using cecal and fecal samples. ↑*Bifidobacteria* and ↑*Lactobacilli* Reduced putrefactive microorganisms.	Not measured	No side effects, such as eructation and bloating, were observed	[107]
Saccharina (Laminiaria) japonica (B)	alginate	4 mg/mL, simulated oral, gastric and small intestinal digestion	Method: 16S rRNA sequencing-based high throughput sequencing, MALDI-TOF/MS. *Bacteroides (Bacteroides finegoldii)*	↑Acetic acid and ↑propionic acid	Specific modulation of *Bacteroides* by alginates	[108]
Eisenia bicyclis (B)	laminarin	2% (*w*/*w*) In vivo in rats	Method: 16S rDNA-DGGE and Pyrosequencing. ↑ *Catabacter hongkongensis* ↑ *Stomatobaculum longum* ↓ *Adlercreuzia* ↓ *Helicobacter*	↓ Indole	Not measured	[109]
Ascophyllum nodosum and *Laminaria japonica*	Fucoidan	0.01% day^{-1} In vivo in mice	Method: 16S rRNA sequencing-based high throughput sequencing. ↑ *Lactobacillus* ↑ *Ruminococcaceae* ↓ *Peptococcus*	Not measured	Reduced the antigen load and the inflammatory response	[110]
Ecklonia radiata (B)	Whole seaweed and polysaccharide fraction (fucoidan and alginate)	5% (*w*/*w*) WS 5% (*w*/*w*) PF in vivo in rats	Method: Real time Q-PCR. ↑*F. prausnitzii* ↑*E. coli* (PF) ↓ *Enterococcus* (WS) ↓*Lactobacillus* ↓*Bifidobacterium* ↓*Firmicutes: Bacteroidetes*	↑ Acetate ↑ Propionate ↑ Butyrate (PF) ↓ Valerate ↓ Hexanoate ↑ Total SCFA ↓ i-Butyrate ↓ i-Valerate ↓ phenol ↓ p-cresol	↑ Butyrate (PF) Linked to anti-inflammatory action	[111]

7. Reduced Use of Antibiotics in Combination with Dietary Seaweeds

Antibiotics have been used as therapeutics for the treatment of animal and human diseases, as prophylactics to prevent infection and as growth promoters in livestock production [112]. Sub-therapeutic levels of antibiotics (i.e., <200 g/ton of feed) have been included in animal diets in order to achieve growth promoting effects (U.S. Food and Drug Administration, 2000). The selection pressure on gut microbes caused by routine use of antibiotics has promoted the development of resistance genes that are capable of horizontal gene transfer between different species of pathogenic bacteria. This unfortunate situation has resulted in the uncontrolled multiplication of resistant bacterial pathogens including *Clostridium, Salmonella,* and *Campylobacter,* which can cause harmful diseases in the host. In addition, alterations in the microbiome within the host gut can lead to a predisposition to infection by other environmental pathogens [113]. In the United States, the Food and Drug Administration's Center for Veterinary Medicine (CVM) has developed a five-year action plan (2018-23) for supporting anti-microbial stewardship in veterinary settings in order to limit or reverse bacterial resistance due to the overuse of antibiotics in food-producing animals. FDA/CVM has set limits on the use of cephalosporin and has withdrawn approval for the use of fluoroquinolones in poultry, because these antibiotics are also commonly used in human medical treatments. In North America and Europe, there is a heightened public awareness of the negative effect of antibiotics in livestock production and an increasing scientific and regulatory interest in developing alternatives to antibiotics [114]. However, in developing countries, the use of antibiotics in animal production is unregulated. This has resulted in sky-rocketing levels of anti-microbial resistance in many jurisdictions and increasingly worldwide.

The growing demand for animal protein in developing countries has resulted in a dramatic increase in the administration of antibiotics to livestock [115]. Since 2000, the demand for protein from meat plateaued in developed countries but grew significantly in developing countries, i.e., 68% in Asia, 64% in Africa and 40% in South America. An increase in animal production has resulted in increased frequencies of infectious disease outbreaks within flocks and tripled the occurrence of antibiotic resistant, zoonotic bacteria (*E. coli, Campylobacter, Salmonella* and *Staphylococcus aureus*). A comparison between developing countries indicated that antibiotic resistance was most widespread in China and India, followed by Brazil and Kenya [115,116].

A recent report from the Infectious Diseases Society of America (IDSA) indicated that there were only ten new drugs in the pipeline (in phase 2 or phase 3 trials) for the treatment of infections caused by pathogens. These drugs, which are under development, might fail to receive regulatory approval by the FDA and are furthermore not guaranteed to be effective against certain antibiotic-resistant pathogens [112].

Potentiating the activity of existing antibiotics using combination therapies could be an alternative strategy to discovering new antibiotics. A range of anti-microbial peptides, molecules, plant extracts and essential oils, all with anti-microbial activity, have demonstrated such combination effects [117,118]. Similarly, seaweeds have been tested in combination with antibiotics to extend the lifespan of fading (off-patent) antibiotics which are utilized in animal production. For example, alginates from certain brown seaweeds have been shown to potentiate the anti-microbial activity of specific antibiotics (i.e., macrolides, β-lactams and tetracyclines) that are efficacious against pathogens such as *Pseudomonas, Acinetobacter* and *Burkholderia* spp. [119]. Functional extracts from the brown seaweeds *Laminaria japonica* and *Sargassum horneri* and the red seaweeds *Gracilaria* sp. and *Porphyra dentata* potentiated the activity of macrolides such as clarithromycin against antibiotic-resistant *E. coli*. Ethanolic extracts of some seaweeds, in combination with clarithromycin, were observed to synergistically inhibit bacterial growth by inhibiting the activity of efflux pumps [120]. Water extracts of two red seaweeds, e.g., *Chondrus crispus* and *Sarcodiotheca gaudichaudii* (SG), in combination with tetracycline and streptomycin, significantly enhanced anti-bacterial activity against *Salmonella* Enteritidis. A water extract from SG at 400 and 800 μg/mL, in combination with sub-lethal concentrations of tetracycline (1 and 1.63 μg/mL), showed complete inhibition of bacterial growth, comparable to full strength tetracycline (23 μg/mL) [121]. The proposed mode of action of the combined effect was the inhibition of quorum sensing in SE *Salmonella*, thereby repressing efflux-related gene expression, resulting in the accumulation of tetracycline within the bacterial cell, ultimately leading to cell death [121]. These findings confirmed the in vitro activities of certain seaweeds and their extracts, which can be employed to increase the lifetime of existing antibiotics. Further research needs to be carried out to test such combinatorial effects in in vivo models such as rats and mice, and then in livestock, to validate these findings. Reduced antibiotic consumption in farm animal production is highly desirable and this may be eventually achieved by feed supplementation of probiotic seaweeds.

8. Commercialization of Various Seaweeds for Animal Feeds

The global commercial market for seaweeds provides a broad range of products for direct or indirect human uses. This was valued at USD 11.48 Billion in 2017, with a CAGR of 8.42% [122]. The growing scope of seaweed-based applications in food, agricultural fertilizers, animal feed additives, pharmaceuticals, cosmetics and personal care is expected to significantly boost market demand. Additionally, rising demands for seaweed-derived hydrocolloids such as agar, alginates and carrageenans also contribute significantly to the total volumes and values of the commercial seaweed market (Agriculture and animal feed applications held the second largest seaweed market share in 2017 [123], and these are anticipated to reach much higher values by 2024, due to the impacts of current R&D (research and development) targeting enhanced animal health and productivity [123]. Table 3 collates information provided by various producers of poultry products with respect to seaweed-based products.

Challenges and Future Prospects

(a) The effect of seaweed harvesting on the environment: The global seaweed industry largely relies on harvesting seaweed as a natural resource. Over-harvesting due to increases in seaweed demand could negatively impact the environment and the sustainability of supply. Science-based management plans to maintain a sustainable cultivation and collection/ harvest strategy for seaweed biomass are critical, particularly since some seaweeds have growth rates which exceed those of many terrestrial crop plants, indicating that the selection and domestication of such seaweeds for cultivation would be an ideal direction for future sustainability [124].

(b) Macroalgal cultivation systems: A sound production strategy is vital to improve supplies of selected seaweed biomass due to the predicted growing market demand over the next 5 years [122]. However, reliable, sustainable and economically viable cultivation of seaweeds represents a major challenge due to the high costs and labor associated with establishing large-scale industrial plants. Current large-scale facilities do not have sufficient capacity to produce the huge quantities of seaweed biomass that are necessary to meet global demand by the animal feed industry [125]. One major roadblock is the inadequate numbers of commercial seaweed farms with on-land tank facilities. One solution would be to establish large-scale production facilities in low-income countries [126]. In 2014, Asian countries collectively produced more than 10 million metric tonnes of cultivated seaweeds, whereas European nations produced comparatively limited quantities (10,000—100,000 metric tonnes) [127]. Integrated multi-trophic aquaculture (IMTA), which involves the co-cultivation of macroalgae with other livestock such as fish and molluscs, could be a viable commercial alternative. This could also create a balanced ecosystem between seaweed crops and aquatic life [128].

Table 3. Summary of major seaweed companies supplying poultry feed and additives.

Business Organization	Product/Source/ Description	Stage of Development/ Operation Level	Product Function/ Claims	Animal Health Sector	Web Address
Ekogea, UK	-BCxF®, prebiotic poultry water additive (comprised of *Ascophyllum Nodosum*). -BCxS, animal housing sanitizer, 100% *Ascophyllum Nodosum*	Commercialized market, industrial scale	-Reduces mortality; -Improves feed conversion, weight gain and overall bird health, gut health; -Reduces ammonia emission and *Campylobacter* levels.	Poultry, pigs and other animals	http://www.ekogea.co.uk/
Ocean Harvest Technology, Ireland	-OceanFeed™ -Ocean Poultry contains a complex blend of seaweeds (red, green and brown) based feed additive	Commercialized market, industrial scale	-Improves body weight gain; -Increases breast meat yield increased.	Poultry, pigs and canines (dogs)	https://www.oceanharvesttechnology.com
Olmix Group	Mycotoxin risk: -MT.X+ -MMLS Digestive efficiency: -MFeed+ -DigestSea Immunity: -Searup -Algimun	Commercialized market, industrial scale	-Reduces mycotoxins in feed; -Immune modulation; -Improves feed efficiency.	Poultry, pigs and canines	https://www.olmix.com
Algea, The Arctic Company, Kristiansund - Omagata	AlgaeFeed 1.4 AlgaeFeed 3.5 Brown seaweed meal containing mineral macro and micro elements	Commercialized market, industrial scale	-Improves animal metabolism and performance; -Improves eggshell quality and production of eggs.	Poultry, pigs, cattle, fish and equines	https://www.algea.com/
FutureFeed, Australia	*Asparagopsis* seaweed-based diet	Conducting trials for commercialization/ pilot scale	-Anti-bacterial properties; -Methane reduction in livestock digestive fermentation.	Livestock	https://www.csiro.au/en/Research/AF/Areas/Food-security/FutureFeed
SeaLac, Seaweed production	*Ascophyllum nodosum*-based seaweed supplement- organic dried seaweed-food grade	Commercialized market, industrial scale	-Improves feed absorption, weight gain in broilers; -Natural alternative to antibiotics; -Improves egg production, egg quality and immune response.	Poultry, pigs and cattle	http://www.sealac.eu/

There is no implied support for any specific product or manufacturer. Information was gathered from a search of web sites in Sept-Nov, 2019.

(c) Heavy metals, mineral, plastic and other safety hazards: Marine algae tend to concentrate heavy metals and other mineral contaminants [129,130]. Seaweeds for food and feed are always tested to measure levels for trace elements (As, Cd, Pb, Sn and Hg) in order to meet national and international regulation and safety standards [126,131]. Other safety hazards for seaweeds may include anti-nutritional factors, radioactive isotopes, ammonium, dioxins and pesticides. In addition, there are reports that seaweeds increasingly contain traces of plastic particles which might affect the utilization of specific seaweeds for human and animal food [132]. Cultivation of specific seaweed species and or their selected cultivars, targeting specific applications, might be necessary in order to guarantee contaminant-free materials [133].

(d) Seasonal variability, harvesting, processing variability: Seasonal variability affects the nutritional profile of seaweeds [133]. Nutritional and biological activities of seaweeds are primarily due to the presence of compounds such as polysaccharides, carotenoids, fatty acids, proteins, peptides, vitamins, minerals and dietary polyphenols. Seaweeds synthesize several of these compounds in response to complex environmental conditions. Thus, the composition of these varies with seasonal variability. Controlling seasonal variability is a major challenge to maintain consistency in the bioactive compounds as nutrients for feed supplement. Effective measures should be implemented for seaweed harvesting and processing in order to maintain consistency in composition of bioactive material [132].

9. Conclusions

This review highlights recent developments in research on selected seaweeds as a valuable and sustainable feed additive for multiple poultry applications. Utilization of selected seaweeds in animal feeds and supplements will improve animal food security and welfare. Advances in scientific evidence from both in vitro and in vivo studies provides promising data to support the utilization of certain seaweeds and their derived compounds to modulate gastrointestinal microbiome and the gut short chain fatty acids (SCFAs).

Dietary polysaccharides from seaweeds are not only a source of anti-microbials but also function as prebiotics and improve the growth of beneficial microflora in gastrointestinal tract. The encouraging data presented in this review supports the need for further research on the use of seaweeds to combat the increasing pressure for an antibiotic-free poultry industry by providing alternatives in the form of natural prebiotics.

Author Contributions: Conceptualization, A.C. and G.K.; Writing—original draft preparation, G.K.; Writing—review and editing, A.C., G.K., B.P., M.T.H.; Supervision, A.C., B.P., M.T.H.; Project administration, A.C. All authors have read and agreed the published version of the manuscript.

Funding: This research received no external funding.

Conflicts of Interest: The authors declare no conflict of interest.

References

1. Guiry, M.D.; Guiry, G.M.; AlgaeBase. World-Wide Electronic Publication, National University of Ireland, Galway. Available online: https://www.algaebase.org (accessed on 6 July 2020).
2. Piconi, P. Edible Seaweed Market Analysis. Available online: http://www.islandinstitute.org/edible-seaweed-market-analysis-2020 (accessed on 25 January 2020).
3. Lorbeer, A.J.; Tham, R.; Zhang, W. Potential products from the highly diverse and endemic macroalgae of Southern Australia and pathways for their sustainable production. *J. Appl. Phycol.* **2013**, *25*, 717–732. [CrossRef]
4. Gupta, S.; Abu-Ghannam, N. Recent developments in the application of seaweeds or seaweed extracts as a means for enhancing the safety and quality attributes of foods. *Innov. Food Sci. Emerg. Technol.* **2011**, *12*, 600–609. [CrossRef]
5. Holdt, S.L.; Kraan, S. Bioactive compounds in seaweed: Functional food applications and legislation. *J. Appl. Phycol.* **2011**, *23*, 543–597. [CrossRef]

6.   FAO. *The State of World Fisheries and Aquaculture 2018–Meeting the Sustainable Development Goals*; FAO: Rome, Italy, 2018.

7.   Comtex. Animal Feed Supplements Market Demand Status with Industry Growth 2020 Latest Trends, Top Manufacturers, Analysis by Market Size and Global Share and Forecast to 2026. Available online: https://www.marketwatch.com/press-release/animal-feed-supplements-market-demand-status-with-industry-growth-2020-latest-trends-top-manufacturers-analysis-by-market-size-and-global-share-and-forecast-to-2026-2020-05-15 (accessed on 25 May 2020).

8.   Allied Market Research. Animal Feed Additives Market by Additive Type (Amino Acids, Antioxidants, Feed Enzymes, Feed Acidifiers, Vitamins, Minerals, Binders, Antibiotics, and Others), Livestock (Swine, Ruminants, Poultry, Aquatic Animals, and Others), Form (Dry, Liquid, and Others), and Function (Single Function and Multifunction): Global Opportunity Analysis and Industry Forecast, 2018–2025. Available online: https://www.alliedmarketresearch.com/animal-feed-additives-market (accessed on 10 September 2019).

9.   Evans, F.D.; Critchley, A.T. Seaweeds for animal production use. *J. Appl. Phycol.* **2014**, *26*, 891–899. [CrossRef]

10. Mišurcová, L. Chemical composition of seaweeds. In *Handbook of Marine Macroalgae: Biotechnology and Applied Phycology*; John Wiley & Sons: West Sussex, UK, 2011; pp. 173–192.

11. Peng, Y.; Hu, J.; Yang, B.; Lin, X.P.; Zhou, X.F.; Yang, X.W.; Liu, Y. Chemical composition of seaweeds. In *Seaweed Sustainability: Food and Non-Food Applications*; Academic Press: Cambridge, MA, USA, 2015; pp. 79–124.

12. Øverland, M.; Mydland, L.T.; Skrede, A. Marine macroalgae as sources of protein and bioactive compounds in feed for monogastric animals. *J. Sci. Food Agric.* **2019**, *99*, 13–24. [CrossRef] [PubMed]

13. Amerah, A.M.; Ravindran, V.; Lentle, R.G.; Thomas, D.G. Feed particle size: Implications on the digestion and performance of poultry. *Worlds Poult. Sci. J.* **2007**, *63*, 439–455. [CrossRef]

14. El-Deek, A.A.; Brikaa, M.A. Nutritional and biological evaluation of marine seaweed as a feedstuff and as a pellet binder in poultry diet. *Int. J. Poult. Sci.* **2009**, *8*, 875–881. [CrossRef]

15. Abudabos, A.M.; Okab, A.B.; Aljumaah, R.S.; Samara, E.M.; Abdoun, K.A.; Al-Haidary, A.A. Nutritional value of green seaweed (Ulva lactuca) for broiler chickens. *Ital. J. Anim. Sci.* **2013**, *12*, e28. [CrossRef]

16. Marković, R.; Šefer, D.; Krstić, M.; Petrujkić, B. Effect of different growth promoters on broiler performance and gut morphology. *Arch. Med. Vet.* **2009**, *41*, 163–169. [CrossRef]

17. Cañedo-Castro, B.; Piñón-Gimate, A.; Carrillo, S.; Ramos, D.; Casas-Valdez, M. Prebiotic effect of Ulva rigida meal on the intestinal integrity and serum cholesterol and triglyceride content in broilers. *J. Appl. Phycol.* **2019**, 1–9. [CrossRef]

18. Choi, Y.J.; Lee, S.R.; Oh, J.W. Effects of dietary fermented seaweed and seaweed fusiforme on growth performance, carcass parameters and immunoglobulin concentration in broiler chicks. *Asian-Australas J. Anim. Sci.* **2014**, *27*, 862–870. [CrossRef] [PubMed]

19. Gumus, R.; Urcar Gelen, S.; Koseoglu, S.; Ozkanlar, S.; Ceylan, Z.G.; Imik, H. The effects of fucoxanthin dietary inclusion on the growth performance, antioxidant metabolism and meat quality of broilers. *Rev. Bras. Cienc. Avic.* **2018**, *20*, 487–496. [CrossRef]

20. Surai, P.F. Selenium in poultry nutrition 1. Antioxidant properties, deficiency and toxicity. *Worlds Poult. Sci. J.* **2002**, *58*, 333–347. [CrossRef]

21. Sobotik, E.; Nelson, J.; Archer, G. The effect of feeding a seaweed extract during heat stress on broiler production and stress. In Proceedings of the Poultry Science Association Annual Meeting, San Antonio, TX, USA, 23–26 July 2018.

22. El-Deek, A.A.; Al-Harthi, M.A.; Abdalla, A.A.; Elbanoby, M.M. The use of brown algae meal in finisher broiler diets. *Egypt. Poult. Sci.* **2011**, *3*, 767–781.

23. Kumar, A.K. Effect of Sargassum wightii on growth, carcass and serum qualities of broiler chickens. *Vet. Sci. Res.* **2018**, *3*, 156.

24. Wiseman, M. Evaluation of Tasco® as A Candidate Prebiotic in Broiler Chickens. Master's Thesis, Dalhousie University, Halifax, NS, Canada, 2012.

25. Walk, C.L.; Addo-Chidie, E.K.; Bedford, M.R.; Adeola, O. Evaluation of a highly soluble calcium source and phytase in the diets of broiler chickens. *Poult. Sci.* **2012**, *91*, 2255–2263. [CrossRef]

26. Bradbury, E.J.; Wilkinson, S.J.; Cronin, G.M.; Walk, C.L.; Cowieson, A.J. The effect of marine calcium source on broiler leg integrity. In Proceedings of the 23rd Annual Australian Poultry Science Symposium, Sydney, Australia, 19–22 February 2012; pp. 85–88.

27. Bradbury, E.J.; Wilkinson, S.J.; Cronin, G.M.; Thomson, P.; Walk, C.L.; Cowieson, A.J. Evaluation of the effect of a highly soluble calcium source in broiler diets supplemented with phytase on performance, nutrient digestibility, foot ash, mobility and leg weakness. *Anim. Prod. Sci.* **2017**, *57*, 2016. [CrossRef]

28. Karimi, S.H. Effects of Red Seaweed (Palmaria palmata) Supplemented Diets Fed to Broiler Chickens Raised under Normal or Stressed Conditions. Master's Thesis, Dalhousie University, Halifax, NS, Canada, 2015.

29. Qadri, S.S.N.; Biswas, A.; Mandal, A.B.; Kumawat, M.; Saxena, R.; Nasir, A.M. Production performance, immune response and carcass traits of broiler chickens fed diet incorporated with Kappaphycus alvarezii. *J. Appl. Phycol.* **2019**, *31*, 753–760. [CrossRef]

30. Balasubramanian, B.; Koo, J.S.; Deun, S.K.; Park, J.H.; Recharla, N.; Park, S.; Kim, I.H. Influence of marine red seaweed supplementation on growth performance, blood metabolites, breast muscle meat quality, fecal consistency score, excreta microbial shedding and noxious gas emission in broilers. In Proceedings of the Poultry Science Association Annual Meeting, San Antonio, TX, USA, 23–26 July 2019.

31. Bach, S.J.; Wang, Y.; McAllister, T.A. Effect of feeding sun-dried seaweed (Ascophyllum nodosum) on fecal shedding of Escherichia coli O157:H7 by feedlot cattle and on growth performance of lambs. *Anim. Feed Sci. Technol.* **2008**, *142*, 17–32. [CrossRef]

32. Michalak, I.; Chojnacka, K.; Dobrzański, Z.; Górecki, H.; Zielińska, A.; Korczyński, M.; Opaliński, S. Effect of macroalgae enriched with microelements on egg quality parameters and mineral content of eggs, eggshell, blood, feathers and droppings. *J. Anim. Physiol. Anim. Nutr.* **2011**, *95*, 374–387. [CrossRef]

33. Wang, S.B.; Jia, Y.H.; Wang, L.H.; Zhu, F.H.; Lin, Y.T. Enteromorpha prolifera supplemental level: Effects on laying performance, egg quality, immune function and microflora in feces of laying hens. *Chin. J. Anim. Nutr.* **2013**, *25*, 1346–1352.

34. Li, Q.; Luo, J.; Wang, C.; Tai, W.; Wang, H.; Zhang, X.; Liu, K.; Jia, Y.; Lyv, X.; Wang, L.; et al. Ulvan extracted from green seaweeds as new natural additives in diets for laying hens. *J. Appl. Phycol.* **2018**, *30*, 2017–2027. [CrossRef]

35. Carrillo, S.; López, E.; Casas, M.M.; Avila, E.; Castillo, R.M.; Carranco, M.E.; Calvo, C.; Pérez-Gil, F. Potential use of seaweeds in the laying hen ration to improve the quality of n-3 fatty acid enriched eggs. *J. Appl. Phycol.* **2008**, *20*, 721–728. [CrossRef]

36. Al-Harthi, M.A.; El-Deek, A.A. Effect of different dietary concentrations of brown marine algae (Sargassum dentifebium) prepared by different methods on plasma and yolk lipid profiles, yolk total carotene and lutein plus zeaxanthin of laying hens. *Ital. J. Anim. Sci.* **2012**, *11*, e64. [CrossRef]

37. Choi, Y.; Lee, E.C.; Na, Y.; Lee, S.R. Effects of dietary supplementation with fermented and non-fermented brown algae by-products on laying performance, egg quality, and blood profile in laying hens. *Asian-Australas. J. Anim. Sci.* **2018**, *31*, 1654–1659. [CrossRef] [PubMed]

38. Stupart, C. Supplementation of Red Seaweed (Chondrus crispus) and Tasco®(Ascophyllum nodosum) in Laying Hen Diets. Master's Thesis, Dalhousie University, Halifax, NS, Canada, 2019.

39. Kulshreshtha, G.; Rathgeber, B.; Stratton, G.; Thomas, N.; Evans, F.; Critchley, A.; Hafting, J.; Prithiviraj, B. Immunology, health, and disease: Feed supplementation with red seaweeds, Chondrus crispus and Sarcodiotheca gaudichaudii, affects performance, egg quality, and gut microbiota of layer hens. *Poult. Sci.* **2014**, *93*. [CrossRef] [PubMed]

40. Abbaspour, B.; Davood, S.S.; Mohammadi-Sangcheshmeh, A. Dietary supplementation of Gracilariopsis persica is associated with some quality related sera and egg yolk parameters in laying quails. *J. Sci. Food Agric.* **2015**, *95*, 643–648. [CrossRef]

41. Mandal, A.B.; Biswas, A.; Mir, N.A.; Tyagi, P.K.; Kapil, D.; Biswas, A.K. Effects of dietary supplementation of Kappaphycus alvarezii on productive performance and egg quality traits of laying hens. *J. Appl. Phycol.* **2019**, *31*, 2065–2072. [CrossRef]

42. Kiarie, E.G.; Mills, A. Role of feed processing on gut health and function in pigs and poultry: Conundrum of optimal particle size and hydrothermal regimens. *Front. Vet. Sci.* **2019**, *6*, 19. [CrossRef] [PubMed]

43. Iqbal, Z.; Drake, K.; Swick, R.A.; Perez-Maldonado, R.A.; Ruhnke, I. Feed particle selection and nutrient intake altered by pecking stone consumption and beak length in free-range laying hens. *Anim. Nutr.* **2019**, *5*, 140–147. [CrossRef]

44. Jacob, J.; Pescatore, T. Avian Digestive System. Available online: http://www2.ca.uky.edu/agcomm/pubs/ASC/ASC203/ASC203.pdf (accessed on 6 November 2019).

45. Missotten, J.A.; Michiels, J.; Dierick, N.; Ovyn, A.; Akbarian, A.; De Smet, S. Effect of fermented moist feed on performance, gut bacteria and gut histo-morphology in broilers. *Br. Poult. Sci.* **2013**, *54*, 627–634. [CrossRef] [PubMed]

46. Xie, P.J.; Huang, L.X.; Zhang, C.H.; Zhang, Y.L. Nutrient assessment of olive leaf residues processed by solid-state fermentation as an innovative feedstuff additive. *J. Appl. Microbiol.* **2016**, *121*, 28–40. [CrossRef] [PubMed]

47. Lin, H.T.V.; Lu, W.J.; Tsai, G.J.; Chou, C.T.; Hsiao, H.I.; Hwang, P.A. Enhanced anti-inflammatory activity of brown seaweed Laminaria japonica by fermentation using Bacillus subtilis. *Process Biochem.* **2016**, *51*, 1945–1953. [CrossRef]

48. Canibe, N.; Jensen, B.B. Fermented liquid feed-Microbial and nutritional aspects and impact on enteric diseases in pigs. *Anim. Feed Sci. Technol.* **2012**, *173*, 17–40. [CrossRef]

49. Carrillo, S.; Ríos, V.H.; Calvo, C.; Carranco, M.E.; Casas, M.; Pérez-Gil, F. N-3 Fatty acid content in eggs laid by hens fed with marine algae and sardine oil and stored at different times and temperatures. *J. Appl. Phycol.* **2012**, *24*, 593–599. [CrossRef]

50. Rizk, Y.S.; Ismail, I.I.; Hafsa, S.H.A.; Eshera, A.A.; Tawfeek, F.A. Effect of dietary green tea and dried seaweed on productive and physiological performance of laying hens during late phase of production. *Egypt. Poult. Sci. J.* **2017**, *37*, 685–706. [CrossRef]

51. Bai, J.; Wang, R.; Yan, L.; Feng, J. Co-supplementation of dietary seaweed powder and antibacterial peptides improves broiler growth performance and immune function. *Braz. J. Poult. Sci.* **2019**, *21*. [CrossRef]

52. Brennan, K.M.; Araujo, L.; Encina, C.; Salvá, B.; Flores, R. Effect of an algae-based antioxidant supplement containing selenium yeast on chicken meat quality. In Proceedings of the Poultry Science Association Annual Meeting, San Antonio, TX, USA, 23–26 July 2013.

53. Bortoluzzi, C.; Schmidt, J.M.; Bordignon, H.L.F.; Fülber, L.M.; Layter, J.R.; Fernandes, J.I.M. Efficacy of yeast derived glucomannan or algae-based antioxidant or both as feed additives to ameliorate mycotoxicosis in heat stressed and unstressed broiler chickens. *Livest. Sci.* **2016**, *193*, 20–25. [CrossRef]

54. Habold, C.; Reichardt, F.; Le Maho, Y.; Angel, F.; Liewig, N.; Lignot, J.H.; Oudart, H. Clay ingestion enhances intestinal triacylglycerol hydrolysis and non-esterified fatty acid absorption. *Br. J. Nutr.* **2009**, *102*, 249–257. [CrossRef]

55. Frobose, H.L.; Erceg, J.A.; Fowler, S.Q.; Tokach, M.D.; DeRouchey, J.M.; Woodworth, J.C.; Dritz, S.S.; Goodband, R.D. The progression of deoxynivalenol-induced growth suppression in nursery pigs and the potential of an algae-modified montmorillonite clay to mitigate these effects. *J. Anim. Sci.* **2016**, *94*, 3746–3759. [CrossRef]

56. Suarez, M.G.; Gallissot, M.; Cierpinski, P. Effect of an algae-clay mix on the use by broiler chickens of a diet containing corn DDGS. In Proceedings of the 4th International Poultry Meat Congress, Antalya, Turkey, 26–30 April 2017; pp. 321–325.

57. Fumonisins, J.; Laurain, M.; Gallissot, M.; Tavares, M.N. Combating the most occurring mycotoxin in South America. In Proceedings of the Poultry Science Association Latin American Annual Meeting, Sao Paulo, Brazil, 6–8 November 2018.

58. Rychen, G.; Aquilina, G.; Azimonti, G.; Bampidis, V.; de Lourdes Bastos, M.; Bories, G.; Chesson, A.; Cocconcelli, P.S.; Flachowsky, G.; Gropp, J.; et al. Safety and efficacy of a preparation of algae interspaced bentonite as a feed additive for all animal species. *EFSA J.* **2016**, *14*. [CrossRef]

59. Ventura, M.R.; Castañon, J.I.R.; McNab, J.M. Nutritional value of seaweed (Ulva rigida) for poultry. *Anim. Feed Sci. Technol.* **1994**, *49*, 87–92. [CrossRef]

60. Cherry, P.; Yadav, S.; Strain, C.R.; Allsopp, P.J.; Mcsorley, E.M.; Ross, R.P.; Stanton, C. Prebiotics from seaweeds: An ocean of opportunity? *Mar. Drugs* **2019**, *17*, 327. [CrossRef] [PubMed]

61. Gudiel-Urbano, M.; Goñi, I. Effect of edible seaweeds (Undaria pinnatifida and Porphyra ternera) on the metabolic activities of intestinal microflora in rats. *Nutr. Res.* **2002**, *22*, 323–331. [CrossRef]

62. Khosravi, M.; Gharibi, D.; Kaviani, F.; Mohammadidust, M. The antibacterial and immunomodulatory effects of carbohydrate fractions of the seaweed gracilaria persica. *J. Med. Microbiol. Infect. Dis.* **2018**, *6*, 57–61. [CrossRef]

63. Ford, L.; Stratakos, A.C.; Theodoridou, K.; Dick, J.T.; Sheldrake, G.N.; Linton, M.; Corcionivoschi, N.; Walsh, P.J. Polyphenols from brown seaweeds as a potential antimicrobial agent in animal feeds. *ACS Omega* **2020**, *5*, 9093–9103. [CrossRef]

64. Alghazeer, R.; Whida, F.; Abduelrhman, E.; Gammoudi, F.; Azwai, S. Screening of antibacterial activity in marine green, red and brown macroalgae from the western coast of Libya. *Nat. Sci.* **2013**, *5*, 7–14. [CrossRef]

65. Wang, W.; Zhang, P.; Yu, G.L.; Li, C.X.; Hao, C.; Qi, X.; Zhang, L.J.; Guan, H.S. Preparation and anti-influenza A virus activity of κ-carrageenan oligosaccharide and its sulphated derivatives. *Food Chem.* **2012**, *133*, 880–888. [CrossRef]

66. Carlucci, M.J.; Scolaro, L.A.; Noseda, M.D.; Cerezo, A.S.; Damonte, E.B. Protective effect of a natural carrageenan on genital herpes simplex virus infection in mice. *Antivir. Res.* **2004**, *64*, 137–141. [CrossRef]

67. Talarico, L.B.; Damonte, E.B. Interference in dengue virus adsorption and uncoating by carrageenans. *Virology* **2007**, *363*, 473–485. [CrossRef] [PubMed]

68. Damonte, E.; Matulewicz, M.; Cerezo, A. Sulfated seaweed polysaccharides as antiviral agents. *Curr. Med. Chem.* **2012**, *11*, 2399–2419. [CrossRef] [PubMed]

69. Hardouin, K.; Bedoux, G.; Burlot, A.S.; Donnay-Moreno, C.; Bergé, J.P.; Nyvall-Collén, P.; Bourgougnon, N. Enzyme-assisted extraction (EAE) for the production of antiviral and antioxidant extracts from the green seaweed Ulva armoricana (Ulvales, Ulvophyceae). *Algal Res.* **2016**, *16*, 233–239. [CrossRef]

70. Trejo-Avila, L.M.; Morales-Martínez, M.E.; Ricque-Marie, D.; Cruz-Suarez, L.E.; Zapata-Benavides, P.; Morán-Santibañez, K.; Rodríguez-Padilla, C. In vitro anti-canine distemper virus activity of fucoidan extracted from the brown alga Cladosiphon okamuranus. *VirusDisease* **2014**, *25*, 474–480. [CrossRef]

71. Elizondo-Gonzalez, R.; Cruz-Suarez, L.E.; Ricque-Marie, D.; Mendoza-Gamboa, E.; Rodriguez-Padilla, C.; Trejo-Avila, L.M. In vitro characterization of the antiviral activity of fucoidan from Cladosiphon okamuranus against Newcastle Disease Virus. *Virol. J.* **2012**, *9*, 307. [CrossRef] [PubMed]

72. Pereira, L.; Critchley, A.T. The COVID 19 novel coronavirus pandemic 2020: Seaweeds to the rescue? Why does substantial, supporting research about the antiviral properties of seaweed polysaccharides seem to go unrecognized by the pharmaceutical community in these desperate times? *J. Appl. Phycol.* **2020**, 1–3. [CrossRef]

73. Leonel pereire antiviral activity of seaweeds and their extracts. In *Therapeutic and Nutritional Uses of Algae*; CRC Press: Boca Raton, FL, USA, 2018; pp. 175–211.

74. Park, H.J.; Kurokawa, M.; Shiraki, K.; Nakamura, N.; Choi, J.S.; Hattori, M. Antiviral activity of the marine alga Symphyocladia latiuscula against herpes simplex virus (HSV-1) in Vitro and its therapeutic efficacy against HSV-1 infection in mice. *Biol. Pharm. Bull.* **2005**, *28*, 2258–2262. [CrossRef]

75. Diogo, J.V.; Novo, S.G.; González, M.J.; Ciancia, M.; Bratanich, A.C. Antiviral activity of lambda-carrageenan prepared from red seaweed (Gigartina skottsbergii) against BoHV-1 and SuHV-1. *Res. Vet. Sci.* **2015**, *98*, 142–144. [CrossRef]

76. Sweeney, T.; Meredith, H.; Ryan, M.T.; Gath, V.; Thornton, K.; O'Doherty, J.V. Effects of Ascophyllum nodosum supplementation on Campylobacter jejuni colonisation, performance and gut health following an experimental challenge in 10 day old chicks. *Innov. Food Sci. Emerg. Technol.* **2016**, *37*, 247–252. [CrossRef]

77. Kulshreshtha, G.; Rathgeber, B.; MacIsaac, J.; Boulianne, M.; Brigitte, L.; Stratton, G.; Thomas, N.A.; Critchley, A.T.; Hafting, J.; Prithiviraj, B. Feed supplementation with red seaweeds, Chondrus crispus and Sarcodiotheca gaudichaudii, reduce Salmonella Enteritidis in laying hens. *Front. Microbiol.* **2017**, *8*. [CrossRef]

78. Song, L.; Chen, X.; Liu, X.; Zhang, F.; Hu, L.; Yue, Y.; Li, K.; Li, P. Characterization and comparison of the structural features, immune-modulatory and anti-avian influenza virus activities conferred by three algal sulfated polysaccharides. *Mar. Drugs* **2016**, *14*, 4. [CrossRef]

79. Aguilar-Briseño, J.A.; Cruz-Suarez, L.E.; Sassi, J.F.; Ricque-Marie, D.; Zapata-Benavides, P.; Mendoza-Gamboa, E.; Rodríguez-Padilla, C.; Trejo-Avila, L.M. Sulphated polysaccharides from Ulva clathrata and Cladosiphon okamuranus seaweeds both inhibit viral attachment/entry and cell-cell fusion, in NDV infection. *Mar. Drugs* **2015**, *13*, 697–712. [CrossRef] [PubMed]

80. Paiva, D.M.; Walk, C.L.; McElroy, A.P. Influence of dietary calcium level, calcium source, and phytase on bird performance and mineral digestibility during a natural necrotic enteritis episode. *Poult. Sci.* **2013**, *92*, 3125–3133. [CrossRef] [PubMed]

81. Ravikumar, S.; Ramanathan, G.; Jacob Inbaneson, S.; Ramu, A. Antiplasmodial activity of two marine polyherbal preparations from Chaetomorpha antennina and Aegiceras corniculatum against Plasmodium falciparum. *Parasitol. Res.* **2011**, *108*, 107–113. [CrossRef]

82. Sampath-Wiley, P.; Neefus, C.D.; Jahnke, L.S. Seasonal effects of sun exposure and emersion on intertidal seaweed physiology: Fluctuations in antioxidant contents, photosynthetic pigments and photosynthetic efficiency in the red alga Porphyra umbilicalis Kützing (Rhodophyta, Bangiales). *J. Exp. Mar. Bio. Ecol.* **2008**, *361*, 83–91. [CrossRef]

83. O'Connor, D.O.; Rubino, J.R. Phenolic compounds. In *Disinfection, Sterilization and Preservation*; Block, S.S., Ed.; Lea and Febiger: Philadelphia, PA, USA, 1991; pp. 204–224.

84. Potin, P.; Bouarab, K.; Küpper, F.; Kloareg, B. Oligosaccharide recognition signals and defence reactions in marine plant-microbe interactions. *Curr. Opin. Microbiol.* **1999**, *2*, 276–283. [CrossRef]

85. Choi, J.; Shin, D.; Kim, M.; Park, J.; Lim, S.; Ryu, S. LsrR-mediated quorum sensing controls invasiveness of salmonella typhimurium by regulating SPI-1 and flagella genes. *PLoS ONE* **2012**, *7*, e37059. [CrossRef] [PubMed]

86. Jesudhasan, P.R.; Cepeda, M.L.; Widmer, K.; Dowd, S.E.; Soni, K.A.; Hume, M.E.; Zhu, J.; Pillai, S.D. Transcriptome analysis of genes controlled by luxS/Autoinducer-2 in salmonella enterica serovar typhimurium. *Foodborne Pathog. Dis.* **2010**, *7*, 399–410. [CrossRef]

87. Janssens, J.C.A.; Steenackers, H.; Robijns, S.; Gellens, E.; Levin, J.; Zhao, H.; Hermans, K.; De Coster, D.; Verhoeven, T.L.; Marchal, K.; et al. Brominated furanones inhibit biofilm formation by Salmonella enterica serovar Typhimurium. *Appl. Environ. Microbiol.* **2008**, *74*, 6639–6648. [CrossRef]

88. Manefield, M.; De Nys, R.; Kumar, N.; Read, R.; Givskov, M.; Steinberg, P.; Kjelleberg, S. Evidence that halogenated furanones from Delisea pulchra inhibit acylated homoserine lactone (AHL)-mediated gene expression by displacing the AHL signal from its receptor protein. *Microbiology* **1999**, *145*, 283–291. [CrossRef]

89. Kulshreshtha, G. The Use of Selected Red Macroalgae (Seaweeds) for the Reduction of *Salmonella* Enteritidis in Poultry. Ph.D. Thesis, Dalhousie University, Halifax, NS, Canada, 2016.

90. Ofek, I.; Mirelman, D.; Sharon, N. Adherence of escherichia coli to human mucosal cells mediated by mannose receptors. *Nature* **1977**, *265*, 623–625. [CrossRef]

91. Karbassi, E.; Asadinezhad, A.; Lehocký, M.; Humpolíček, P.; Vesel, A.; Novák, I.; Sáha, P. Antibacterial performance of alginic acid coating on polyethylene film. *Int. J. Mol. Sci.* **2014**, *15*, 14684–14696. [CrossRef]

92. Kulshreshtha, G.; Borza, T.; Rathgeber, B.; Stratton, G.S.; Thomas, N.A.; Critchley, A.; Hafting, J.; Prithiviraj, B. Red seaweeds Sarcodiotheca gaudichaudii and chondrus crispus down regulate virulence factors of salmonella enteritidis and induce immune responses in Caenorhabditis elegans. *Front. Microbiol.* **2016**, *7*. [CrossRef]

93. Hirayama, M.; Shibata, H.; Imamura, K.; Sakaguchi, T.; Hori, K. High-mannose specific lectin and its recombinants from a carrageenophyta kappaphycus alvarezii represent a potent anti-HIV activity through high-affinity binding to the viral envelope glycoprotein gp120. *Mar. Biotechnol.* **2016**, *18*, 215–231. [CrossRef]

94. Zhang, W.; Oda, T.; Yu, Q.; Jin, J.-O. Fucoidan from macrocystis pyrifera has powerful immune-modulatory effects compared to three other fucoidans. *Mar. Drugs* **2015**, *13*, 1084–1104. [CrossRef] [PubMed]

95. Shi, Q.; Wang, A.; Lu, Z.; Qin, C.; Hu, J.; Yin, J. Overview on the antiviral activities and mechanisms of marine polysaccharides from seaweeds. *Carbohydr. Res.* **2017**, *453–454*, 1–9. [CrossRef]

96. Patterson, J.A.; Burkholder, K. Application of prebiotics and probiotics in poultry production. *Poult. Sci.* **2003**, *82*, 627–631. [CrossRef]

97. Hajati, H.; Rezaei, M. The application of prebiotics in poultry production mitochondrial energetics in response to different fat types in heat-stressed broilers view project effects of fat sources on energetics and redox statues of liver mitochondria in heat-stressd broilers Vi. *Artic. Int. J. Poult. Sci.* **2010**, *9*, 298–304. [CrossRef]

98. Gibson, G.R.; Roberfroid, M.B. Dietary modulation of the human colonic microbiota: Introducing the concept of prebiotics. *J. Nutr.* **1995**, *125*, 1401–1412. [CrossRef] [PubMed]

99. Ricke, S.C. Impact of prebiotics on poultry production and food safety. *Yale J. Biol Med.* **2018**, *91*, 151–159. [PubMed]

100. Cummings, J.H.; Macfarlane, G.T. Gastrointestinal effects of prebiotics. *Br. J. Nutr.* **2002**, *87*, S145–S151. [CrossRef]

101. Reddy, B.S. Possible mechanisms by which pro- and prebiotics influence colon carcinogenesis and tumor growth. *J. Nutr.* **1999**, *129*, 1478S–1482S. [CrossRef] [PubMed]

102. Mussatto, S.I.; Mancilha, I.M. Non-digestible oligosaccharides: A review. *Carbohydr. Polym.* **2007**, *68*, 587–597. [CrossRef]

103. Apajalahti, J. Comparative gut microflora, metabolic challenges, and potential opportunities. *J. Appl. Poult. Res.* **2005**, *14*, 444–453. [CrossRef]

104. Zheng, L.X.; Chen, X.Q.; Cheong, K.L. Current trends in marine algae polysaccharides: The digestive tract, microbial catabolism, and prebiotic potential. *Int. J. Biol. Macromol.* **2020**, *15*, 344–354. [CrossRef]

105. Liu, J.; Kandasamy, S.; Zhang, J.; Kirby, C.W.; Karakach, T.; Hafting, J.; Critchley, A.T.; Evans, F.; Prithiviraj, B. Prebiotic effects of diet supplemented with the cultivated red seaweed Chondrus crispus or with fructo-oligo-saccharide on host immunity, colonic microbiota and gut microbial metabolites. *BMC Complement. Altern. Med.* **2015**, *15*, 279. [CrossRef]

106. Ramnani, P.; Chitarrari, R.; Tuohy, K.; Grant, J.; Hotchkiss, S.; Philp, K.; Campbell, R.; Gill, C.; Rowland, I. Invitro fermentation and prebiotic potential of novel low molecular weight polysaccharides derived from agar and alginate seaweeds. *Anaerobe* **2012**, *18*, 1–6. [CrossRef]

107. Hu, B.; Gong, Q.; Wang, Y.; Ma, Y.; Li, J.; Yu, W. Prebiotic effects of neoagaro-oligosaccharides prepared by enzymatic hydrolysis of agarose. *Anaerobe* **2006**, *12*, 260–266. [CrossRef]

108. Ai, C.; Jiang, P.; Liu, Y.; Duan, M.; Sun, X.; Luo, T.; Jiang, G.; Song, S. The specific use of alginate from: Laminaria japonica by Bacteroides species determined its modulation of the Bacteroides community. *Food Funct.* **2019**, *10*, 4304–4314. [CrossRef]

109. Nakata, T.; Kyoui, D.; Takahashi, H.; Kimura, B.; Kuda, T. Inhibitory effects of laminaran and alginate on production of putrefactive compounds from soy protein by intestinal microbiota in vitro and in rats. *Carbohydr. Polym.* **2016**, *143*, 61–69. [CrossRef]

110. Shang, Q.; Shan, X.; Cai, C.; Hao, J.; Li, G.; Yu, G. Dietary fucoidan modulates the gut microbiota in mice by increasing the abundance of: Lactobacillus and Ruminococcaceae. *Food Funct.* **2016**, *7*, 3224–3232. [CrossRef]

111. Charoensiddhi, S.; Conlon, M.A.; Methacanon, P.; Franco, C.M.M.; Su, P.; Zhang, W. Gut health benefits of brown seaweed Ecklonia radiata and its polysaccharides demonstrated in vivo in a rat model. *J. Funct. Foods* **2017**, *37*, 676–684. [CrossRef]

112. Casewell, M.; Friis, C.; Marco, E.; Mcmullin, P.; Phillips, I. The European ban on growth-promoting antibiotics and emerging consequences for human and animal health. *J. Antimicrob. Chemother.* **2003**, *52*, 159–161. [CrossRef] [PubMed]

113. Pickard, J.M.; Zeng, M.Y.; Caruso, R.; Núñez, G. Gut microbiota: Role in pathogen colonization, immune responses, and inflammatory disease. *Immunol. Rev.* **2017**, *279*, 70–89. [CrossRef]

114. U.S. Food and Drug Administration. Antimicrobial Resistance. Available online: https://www.fda.gov/animal-veterinary/safety-health/antimicrobial-resistance (accessed on 25 May 2020).

115. Maday, J. Princeton Study: Antibiotic Resistance Increasing Globally. Available online: https://www.drovers.com/article/princeton-study-antibiotic-resistance-increasing-globally (accessed on 16 November 2019).

116. Van Boeckel, T.P.; Pires, J.; Silvester, R.; Zhao, C.; Song, J.; Criscuolo, N.G.; Gilbert, M.; Bonhoeffer, S.; Laxminarayan, R. Global trends in antimicrobial resistance in animals in low- And middle-income countries. *Science* **2019**, *365*. [CrossRef]

117. Hussin, W.A.; El-Sayed, W.M. Synergic interactions between selected botanical extracts and tetracycline against gram positive and gram negative bacteria. *J. Biol. Sci.* **2011**, *11*, 433–441. [CrossRef]

118. Chovanová, R.; Mezovská, J.; Vaverková, Š.; Mikulášová, M. The inhibition the Tet(K) efflux pump of tetracycline resistant Staphylococcus epidermidis by essential oils from three Salvia species. *Lett. Appl. Microbiol.* **2015**, *61*, 58–62. [CrossRef]

119. Khan, S.; Tøndervik, A.; Sletta, H.; Klinkenberg, G.; Emanuel, C.; Onsøyen, E.; Myrvold, R.; Howe, R.A.; Walsh, T.R.; Hill, K.E.; et al. Overcoming drug resistance with alginate oligosaccharides able to potentiate the action of selected antibiotics. *Antimicrob. Agents Chemother.* **2012**, *56*, 5134–5141. [CrossRef] [PubMed]

120. Lu, W.-J.; Lin, H.-J.; Hsu, P.-H.; Lai, M.; Chiu, J.-Y.; Lin, H.-T.V. Brown and red seaweeds serve as potential efflux pump inhibitors for drug-resistant escherichia coli. *Evid.–Based Complement. Altern. Med.* **2019**. [CrossRef]

121. Prithiviraj, B.; Kulshreshtha, G. Use of Floridoside or Isethionic Acid to Potentiate Antimicrobial Activity of Antibiotics. Available online: https://patents.google.com/patent/US20190038651A1/en (accessed on 20 January 2020).

122. Fortune Business Insight. Commercial Seaweed Market Size, Share and Global Trend by Product Type (Red, Brown, Green), by Form (Flakes, Powder, Liquid), by END user (Food and Beverage, Agriculture Fertilizer, Animal Feed Additives, Pharmaceuticals, Cosmetics, and Personal Care) and Regional Forecast Till 2025. Available online: http://web.archive.org/web/20200510003328/https://www.fortunebusinessinsights.com/industry-reports/commercial-seaweed-market-100077 (accessed on 24 September 2019).

123. Grand View Research Inc. Commercial Seaweeds Market Size, Share & Trends Analysis Report By Product (Brown Seaweeds, Red Seaweeds, Green Seaweeds), By Form (Liquid, Powdered, Flakes), By Application, By Region, And Segment Forecasts, 2020–2027. Available online: https://www.grandviewresearch.com/industry-analysis/commercial-seaweed-market (accessed on 25 May 2020).

124. Mac Monagail, M.; Cornish, L.; Morrison, L.; Araújo, R.; Critchley, A.T. Sustainable harvesting of wild seaweed resources. *Eur. J. Phycol.* **2017**, *52*, 371–390. [CrossRef]

125. Cornish, L. The animal kingdom and seaweeds. *J. Mar. Sci. Pers. Comm.* **2020**. in preparation.

126. Lerat, Y.; Cornish, M.L.; Critchley, A.T. Applications of algal biomass in global food and feed markets: From traditional usage to the potential for functional products. In *Blue Biotechnology: Production and Use of Marine Molecules*; Wiley YHC Wiley-VCH Verlag GmbH & Co. KGaA: Weinheim, Germany, 2018; pp. 143–189.

127. Buschmann, A.H.; Camus, C.; Infante, J.; Neori, A.; Israel, Á.; Hernández-González, M.C.; Pereda, S.V.; Gomez-Pinchetti, J.L.; Golberg, A.; Tadmor-Shalev, N.; et al. Seaweed production: Overview of the global state of exploitation, farming and emerging research activity. *Eur. J. Phycol.* **2017**, *52*, 391–406. [CrossRef]

128. Ratcliff, J.J.; Wan, A.H.L.; Edwards, M.D.; Soler-Vila, A.; Johnson, M.P.; Abreu, M.H.; Morrison, L. Metal content of kelp (Laminaria digitata) co-cultivated with Atlantic salmon in an Integrated Multi-Trophic Aquaculture system. *Aquaculture* **2016**, *450*, 234–243. [CrossRef]

129. Feng, Z.; Zhang, T.; Shi, H.; Gao, K.; Huang, W.; Xu, J.; Wang, J.; Wang, R.; Li, J.; Gao, G. Microplastics in bloom-forming macroalgae: Distribution, characteristics and impacts. *J. Hazard. Mater.* **2020**, *397*, 122752. [CrossRef] [PubMed]

130. Feng, Z.; Zhang, T.; Wang, J.; Huang, W.; Wang, R.; Xu, J.; Fu, G.; Gao, G. Spatio-temporal features of microplastics pollution in macroalgae growing in an important mariculture area, China. *Sci. Total Environ.* **2020**, *719*, 137490. [CrossRef]

131. Mac Monagail, M.; Morrison, L. Arsenic speciation in a variety of seaweeds and associated food products. *Compr. Anal. Chem.* **2019**, *85*, 267–310.

132. Rajauria, G. Seaweeds: A sustainable feed source for livestock and aquaculture. *Seaweed Sustain.* **2015**, 389–420.

133. Stengel, D.B.; Connan, S.; Popper, Z.A. Algal chemodiversity and bioactivity: Sources of natural variability and implications for commercial application. *Biotechnol. Adv.* **2011**, *29*, 483–501. [CrossRef]

MDPI
St. Alban-Anlage 66
4052 Basel
Switzerland
Tel. +41 61 683 77 34
Fax +41 61 302 89 18
www.mdpi.com

Journal of Marine Science and Engineering Editorial Office
E-mail: jmse@mdpi.com
www.mdpi.com/journal/jmse

www.ingramcontent.com/pod-product-compliance
Lightning Source LLC
LaVergne TN
LVHW070648170726
843515LV00004B/349